Dimensions Math®
Teacher's Guide 1A

Authors and Reviewers

Cassandra Turner

Allison Coates

Jenny Kempe

Bill Jackson

Tricia Salerno

Singapore Math Inc.

Published by Singapore Math Inc.

19535 SW 129th Avenue
Tualatin, OR 97062
www.singaporemath.com

Dimensions Math® Teacher's Guide 1A
ISBN 978-1-947226-32-6

First published 2018
Reprinted 2019, 2020

Printed in China

Acknowledgments

Editing by the Singapore Math Inc. team.
Design and illustration by Cameron Wray with Carli Fronius.

Teacher's Guide 1A

Contents

Chapter		Lesson	Page

Chapter	Lesson	Page

Chapter	Lesson	Page

Dimensions Math® Curriculum

The **Dimensions Math®** series is a Pre-Kindergarten to Grade 5 series based on the pedagogy and methodology of math education in Singapore. The main goal of the **Dimensions Math®** series is to help students develop competence and confidence in mathematics.

The series follows the principles outlined in the Singapore Mathematics Framework below.

Pedagogical Approach and Methodology

- Through Concrete-Pictorial-Abstract development, students view the same concepts over time with increasing levels of abstraction.
- Thoughtful sequencing creates a sense of continuity. The content of each grade level builds on that of preceding grade levels. Similarly, lessons build on previous lessons within each grade.
- Group discussion of solution methods encourages expansive thinking.
- Interesting problems and activities provide varied opportunities to explore and apply skills.
- Hands-on tasks and sharing establish a culture of collaboration.
- Extra practice and extension activities encourage students to persevere through challenging problems.
- Variation in pictorial representation (number bonds, bar models, etc.) and concrete representation (straws, linking cubes, base ten blocks, discs, etc.) broaden student understanding.

Each topic is introduced, then thoughtfully developed through the use of a variety of learning experiences, problem solving, student discourse, and opportunities for mastery of skills. This combination of hands-on practice, in-depth exploration of topics, and mathematical variability in teaching methodology allows students to truly master mathematical concepts.

Singapore Mathematics Framework

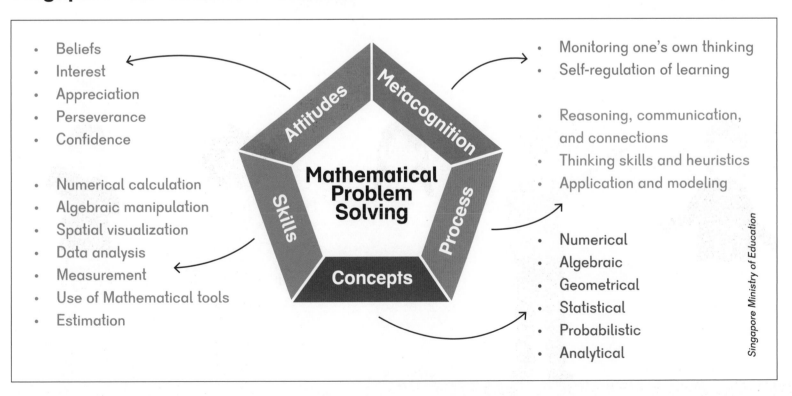

- Beliefs
- Interest
- Appreciation
- Perseverance
- Confidence

- Numerical calculation
- Algebraic manipulation
- Spatial visualization
- Data analysis
- Measurement
- Use of Mathematical tools
- Estimation

Attitudes

Metacognition

Skills

Mathematical Problem Solving

Process

Concepts

- Monitoring one's own thinking
- Self-regulation of learning

- Reasoning, communication, and connections
- Thinking skills and heuristics
- Application and modeling

- Numerical
- Algebraic
- Geometrical
- Statistical
- Probabilistic
- Analytical

Singapore Ministry of Education

Dimensions Math® Program Materials

Textbooks

Textbooks are designed to help students build a solid foundation in mathematical thinking and efficient problem solving. Careful sequencing of topics, well-chosen problems, and simple graphics foster deep conceptual understanding and confidence. Mental math, problem solving, and correct computation are given balanced attention in all grades. As skills are mastered, students move to increasingly sophisticated concepts within and across grade levels.

Students work through the textbook lessons with the help of five friends: Emma, Alex, Sofia, Dion, and Mei. The characters appear throughout the series and help students develop metacognitive reasoning through questions, hints, and ideas.

A pencil icon ▭▶ at the end of the textbook lessons links to exercises in the workbooks.

Workbooks

Workbooks provide additional problems that range from basic to challenging. These allow students to independently review and practice the skills they have learned.

Teacher's Guides

Teacher's Guides include lesson plans, mathematical background, games, helpful suggestions, and comprehensive resources for daily lessons.

Tests

Tests contain differentiated assessments to systematically evaluate student progress.

Emma   Alex   Sofia   Dion   Mei

Online Resources

The following can be downloaded from dimensionsmath.com.

- **Blackline Masters** used for various hands-on tasks.

- **Letters Home** to be emailed or sent home with students for continued exploration. These outline what the student is learning in math class and offer suggestions for related activities at home. Reinforcement at home supports deep understanding of mathematical concepts.

- **Material Lists** for each chapter and lesson, so teachers and classroom helpers can prepare ahead of time.

- **Activities** that can done with students who need more practice or a greater challenge, organized by concept, chapter, and lesson.

- **Standards Alignments** for various states.

Using the Teacher's Guide

This guide is designed to assist in planning daily lessons. It should be considered a helping hand between the curriculum and the classroom. It provides introductory notes on mathematical content, key points, and suggestions for activities. It also includes ideas for differentiation within each lesson, and answers and solutions to textbook and workbook problems.

Each chapter of the guide begins with the following.

● Overview

Includes objectives and suggested number of class periods for each chapter.

● Notes

Highlights key learning points, provides background on math concepts, explains the purpose of certain activities, and helps teachers understand the flow of topics throughout the year.

● Materials

Lists materials, manipulatives, and Blackline Masters used in the Think and Learn sections of the guide. It also includes suggested storybooks. Many common classroom manipulatives are used throughout the curriculum. When a lesson refers to a whiteboard and markers, any writing materials can be used. Blackline Masters can be found at dimensionsmath.com.

The guide goes through the Chapter Openers, Daily Lessons, and Practices of each chapter, and cumulative reviews in the following general format.

● <u>Chapter Opener</u>

Provides talking points for discussion to prepare students for the math concepts to be introduced.

● <u>Think</u>

Offers structure for teachers to guide student inquiry. Provides various methods and activities to solve initial textbook problems or tasks.

● <u>Learn</u>

Guides teachers to analyze student methods from Think to arrive at the main concepts of the lesson through discussion and study of the pictorial representations in the textbook.

● <u>Do</u>

Expands on specific problems with strategies, additional practice, and remediation.

● <u>Activities</u>

Allows students to practice concepts through individual, small group, and whole group hands-on tasks and games, including suggestions for outdoor play (most of which can be modified for a gymnasium or classroom).

Level of difficulty in the games and activities are denoted by the following symbols.

- ● Foundational activities
- ▲ On-level activities
- ★ Challenge or extension activities

● <u>Brain Works</u>

Provides opportunities for students to extend their mathematical thinking.

Discussion is a critical component of each lesson. Teachers are encouraged to let students discuss their reasoning. As each classroom is different, this guide does not anticipate all situations. The following questions can help students articulate their thinking and increase their mastery:

- Why? How do you know?
- Can you explain that?
- Can you draw a picture of that?
- Is your answer reasonable? How do you know?
- How is this task like the one we did before? How is it different?
- What is alike and what is different about…?
- Can you solve that a different way?
- Yes! You're right! How do you know it's true?
- What did you learn before that can help you solve this problem?
- Can you summarize what your classmate shared?
- What conclusion can you draw from the data?

Each lesson is designed to take one day. If your calendar allows, you may choose to spend more than one day on certain lessons. Throughout the guide, there are notes to extend on learning activities to make them more challenging. Lesson structures and activities do not have to conform exactly to what is shown in the guide. Teachers are encouraged to exercise their discretion in using this material in a way that best suits their classes.

Textbooks are designed to last multiple years. Textbook problems with a ▮ (or a blank line for terms) are meant to invite active participation.

Dimensions Math® Scope & Sequence

PKA

Chapter 1
Match, Sort, and Classify

Red and Blue
Yellow and Green
Color Review
Soft and Hard
Rough, Bumpy, and Smooth
Sticky and Grainy
Size — Part 1
Size — Part 2
Sort Into Two Groups
Practice

Chapter 2
Compare Objects

Big and Small
Long and Short
Tall and Short
Heavy and Light
Practice

Chapter 3
Patterns

Movement Patterns
Sound Patterns
Create Patterns
Practice

Chapter 4
Numbers to 5 — Part 1

Count 1 to 5 — Part 1
Count 1 to 5 — Part 2
Count Back

Count On and Back
Count 1 Object
Count 2 Objects
Count Up to 3 Objects
Count Up to 4 Objects
Count Up to 5 Objects
How Many? — Part 1
How Many? — Part 2
How Many Now? — Part 1
How Many Now? — Part 2
Practice

Chapter 5
Numbers to 5 — Part 2

1, 2, 3
1, 2, 3, 4, 5 — Part 1
1, 2, 3, 4, 5 — Part 2
How Many? — Part 1
How Many? — Part 2
How Many Do You See?
How Many Do You See Now?
Practice

Chapter 6
Numbers to 10 — Part 1

0
Count to 10 — Part 1
Count to 10 — Part 2
Count Back
Order Numbers
Count Up to 6 Objects
Count Up to 7 Objects
Count Up to 8 Objects
Count Up to 9 Objects
Count Up to 10 Objects
    — Part 1

Count Up to 10 Objects
    — Part 2
How Many?
Practice

Chapter 7
Numbers to 10 — Part 2

6
7
8
9
10
0 to 10
Count and Match — Part 1
Count and Match — Part 2
Practice

PKB

Chapter 8
Ordinal Numbers

First
Second and Third
Fourth and Fifth
Practice

Chapter 9
Shapes and Solids

Cubes, Cylinders, and Spheres
Cubes
Positions
Build with Solids
Rectangles and Circles
Squares
Triangles

Dimensions Math® Scope & Sequence

Count Up to 10 Things —
    Part 2
Recognize the Numbers
    6 to 10
Write the Numbers 6 and 7
Write the Numbers 8, 9,
    and 10
Write the Numbers 6 to 10
Count and Write the
    Numbers 1 to 10
Ordinal Positions
One More Than
Practice

Chapter 4
Shapes and Solids

Curved or Flat
Solid Shapes
Closed Shapes
Rectangles
Squares
Circles and Triangles
Where is It?
Hexagons
Sizes and Shapes
Combine Shapes
Graphs
Practice

Chapter 5
Compare Height, Length, Weight, and Capacity

Comparing Height
Comparing Length
Height and Length — Part 1
Height and Length — Part 2
Weight — Part 1

Weight — Part 2
Weight — Part 3
Capacity — Part 1
Capacity — Part 2
Practice

Chapter 6
Comparing Numbers Within 10

Same and More
More and Fewer
More and Less
Practice — Part 1
Practice — Part 2

KB

Chapter 7
Numbers to 20

Ten and Some More
Count Ten and Some More
Two Ways to Count
Numbers 16 to 20
Number Words 0 to 10
Number Words 11 to 15
Number Words 16 to 20
Number Order
1 More Than or Less Than
Practice — Part 1
Practice — Part 2

Chapter 8
Number Bonds

Putting Numbers Together
    — Part 1

Putting Numbers Together
    — Part 2
Parts Making a Whole
Look for a Part
Number Bonds for 2, 3, and 4
Number Bonds for 5
Number Bonds for 6
Number Bonds for 7
Number Bonds for 8
Number Bonds for 9
Number Bonds for 10
Practice — Part 1
Practice — Part 2
Practice — Part 3

Chapter 9
Addition

Introduction to Addition —
    Part 1
Introduction to Addition —
    Part 2
Introduction to Addition —
    Part 3
Addition
Count On — Part 1
Count On — Part 2
Add Up to 3 and 4
Add Up to 5 and 6
Add Up to 7 and 8
Add Up to 9 and 10
Addition Practice
Practice

Chapter 10
Subtraction

Take Away to Subtract —
    Part 1

Dimensions Math® Scope & Sequence

Compare Numbers to 20
Addition
Subtraction
Practice

Chapter 6
Addition to 20

Add by Making 10 — Part 1
Add by Making 10 — Part 2
Add by Making 10 — Part 3
Addition Facts to 20
Practice

Chapter 7
Subtraction Within 20

Subtract from 10 — Part 1
Subtract from 10 — Part 2
Subtract the Ones First
Word Problems
Subtraction Facts Within 20
Practice

Chapter 8
Shapes

Solid and Flat Shapes
Grouping Shapes
Making Shapes
Practice

Chapter 9
Ordinal Numbers

Naming Positions
Word Problems
Practice
Review 2

1B

Chapter 10
Length

Comparing Lengths Directly
Comparing Lengths Indirectly
Comparing Lengths with Units
Practice

Chapter 11
Comparing

Subtraction as Comparison
Making Comparison
    Subtraction Stories
Picture Graphs
Practice

Chapter 12
Numbers to 40

Numbers to 40
Tens and Ones
Counting by Tens and Ones
Comparing
Practice

Chapter 13
Addition and Subtraction Within 40

Add Ones
Subtract Ones
Make the Next Ten
Use Addition Facts
Subtract from Tens
Use Subtraction Facts
Add Three Numbers
Practice

Chapter 14
Grouping and Sharing

Adding Equal Groups
Sharing
Grouping
Practice

Chapter 15
Fractions

Halves
Fourths
Practice
Review 3

Chapter 16
Numbers to 100

Numbers to 100
Tens and Ones
Count by Ones or Tens
Compare Numbers to 100
Practice

Chapter 17
Addition and Subtraction Within 100

Add Ones — Part 1
Add Tens
Add Ones — Part 2
Add Tens and Ones — Part 1
Add Tens and Ones — Part 2
Subtract Ones — Part 1
Subtract from Tens
Subtract Ones — Part 2
Subtract Tens

Dimensions Math® Scope & Sequence

Dividing by 5 and 10
Practice C
Word Problems
Review 2

2B

Chapter 8
Mental Calculation

Adding Ones Mentally
Adding Tens Mentally
Making 100
Adding 97, 98, or 99
Practice A
Subtracting Ones Mentally
Subtracting Tens Mentally
Subtracting 97, 98, or 99
Practice B
Practice C

Chapter 9
Multiplication and Division of 3 and 4

The Multiplication Table of 3
Multiplication Facts of 3
Dividing by 3
Practice A
The Multiplication Table of 4
Multiplication Facts of 4
Dividing by 4
Practice B
Practice C

Chapter 10
Money

Making $1
Dollars and Cents
Making Change
Comparing Money
Practice A
Adding Money
Subtracting Money
Practice B

Chapter 11
Fractions

Halves and Fourths
Writing Unit Fractions
Writing Fractions
Fractions that Make 1 Whole
Comparing and Ordering Fractions
Practice
Review 3

Chapter 12
Time

Telling Time
Time Intervals
A.M. and P.M.
Practice

Chapter 13
Capacity

Comparing Capacity
Units of Capacity
Practice

Chapter 14
Graphs

Picture Graphs
Bar Graphs
Practice

Chapter 15
Shapes

Straight and Curved Sides
Polygons
Semicircles and Quarter-circles
Patterns
Solid Shapes
Practice
Review 4
Review 5

3A

Chapter 1
Numbers to 10,000

Numbers to 10,000
Place Value — Part 1
Place Value — Part 2
Comparing Numbers
The Number Line
Practice A
Number Patterns
Rounding to the Nearest Thousand
Rounding to the Nearest Hundred
Rounding to the Nearest Ten
Practice B

Dimensions Math® Scope & Sequence

The Multiplication Table of 9
Multiplying by 8 and 9
Dividing by 8 and 9
Practice B

Chapter 9
Fractions — Part 1

Fractions of a Whole
Fractions on a Number Line
Comparing Fractions with
    Like Denominators
Comparing Fractions with
    Like Numerators
Practice

Chapter 10
Fractions — Part 2

Equivalent Fractions
Finding Equivalent Fractions
Simplifying Fractions
Comparing Fractions — Part 1
Comparing Fractions — Part 2
Practice A
Adding and Subtracting
    Fractions — Part 1
Adding and Subtracting
    Fractions — Part 2
Practice B

Chapter 11
Measurement

Meters and Centimeters
Subtracting from Meters
Kilometers
Subtracting from Kilometers
Liters and Milliliters
Kilograms and Grams

Word Problems
Practice
Review 3

Chapter 12
Geometry

Circles
Angles
Right Angles
Triangles
Properties of Triangles
Properties of Quadrilaterals
Using a Compass
Practice

Chapter 13
Area and Perimeter

Area
Units of Area
Area of Rectangles
Area of Composite Figures
Practice A
Perimeter
Perimeter of Rectangles
Area and Perimeter
Practice B

Chapter 14
Time

Units of Time
Calculating Time — Part 1
Practice A
Calculating Time — Part 2
Calculating Time — Part 3
Calculating Time — Part 4
Practice B

Chapter 15
Money

Dollars and Cents
Making $10
Adding Money
Subtracting Money
Word Problems
Practice
Review 4
Review 5

4A

Chapter 1
Numbers to One Million

Numbers to 100,000
Numbers to 1,000,000
Number Patterns
Comparing and Ordering
    Numbers
Rounding 5-Digit Numbers
Rounding 6-Digit Numbers
Calculations and Place Value
Practice

Chapter 2
Addition and Subtraction

Addition
Subtraction
Other Ways to Add and
    Subtract — Part 1
Other Ways to Add and
    Subtract — Part 2
Word Problems

Dimensions Math® Scope & Sequence

Chapter 11
Area and Perimeter

Area of Rectangles — Part 1
Area of Rectangles — Part 2
Area of Composite Figures
Perimeter — Part 1
Perimeter — Part 2
Practice

Chapter 12
Decimals

Tenths — Part 1
Tenths — Part 2
Hundredths — Part 1
Hundredths — Part 2
Expressing Decimals as
    Fractions in Simplest Form
Expressing Fractions as
    Decimals
Practice A
Comparing and Ordering
    Decimals
Rounding Decimals
Practice B

Chapter 13
Addition and Subtraction of Decimals

Adding and Subtracting Tenths
Adding Tenths with Regrouping
Subtracting Tenths with
    Regrouping
Practice A
Adding Hundredths
Subtracting from 1 and 0.1
Subtracting Hundredths
Money, Decimals, and Fractions

Practice B
Review 3

Chapter 14
Multiplication and Division of Decimals

Multiplying Tenths and
    Hundredths
Multiplying Decimals by a
    Whole Number — Part 1
Multiplying Decimals by a
    Whole Number — Part 2
Practice A
Dividing Tenths and Hundredths
Dividing Decimals by a Whole
    Number — Part 1
Dividing Decimals by a Whole
    Number — Part 2
Dividing Decimals by a Whole
    Number — Part 3
Practice B

Chapter 15
Angles

The Size of Angles
Measuring Angles
Drawing Angles
Adding and Subtracting Angles
Reflex Angles
Practice

Chapter 16
Lines and Shapes

Perpendicular Lines
Parallel Lines
Drawing Perpendicular and
    Parallel Lines
Quadrilaterals

Lines of Symmetry
Symmetrical Figures and
    Patterns
Practice

Chapter 17
Properties of Cuboids

Cuboids
Nets of Cuboids
Faces and Edges of Cuboids
Practice
Review 4
Review 5

5A

Chapter 1
Whole Numbers

Numbers to One Billion
Multiplying by 10, 100, and
    1,000
Dividing by 10, 100, and 1,000
Multiplying by Tens,
    Hundreds, and Thousands
Dividing by Tens, Hundreds,
    and Thousands
Practice

Chapter 2
Writing and Evaluating Expressions

Expressions with Parentheses
Order of Operations — Part 1
Order of Operations — Part 2

Dimensions Math® Scope & Sequence

Conversion of Measures
Mental Calculation
Practice B

Chapter 10
The Four Operations of Decimals

Adding Decimals to
    Thousandths
Subtracting Decimals
Multiplying by 0.1 or 0.01
Multiplying by a Decimal
Practice A
Dividing by a Whole Number
    — Part 1
Dividing by a Whole Number
    — Part 2
Dividing a Whole Number by
    0.1 and 0.01
Dividing a Whole Number by
    a Decimal
Practice B

Chapter 11
Geometry

Measuring Angles
Angles and Lines
Classifying Triangles
The Sum of the Angles in a
    Triangle
The Exterior Angle of a
    Triangle
Classifying Quadrilaterals
Angles of Quadrilaterals
    — Part 1
Angles of Quadrilaterals
    — Part 1

Drawing Triangles and
    Quadrilaterals
Practice

Chapter 12
Data Analysis and Graphs

Average — Part 1
Average — Part 2
Line Plots
Coordinate Graphs
Line Graphs
Practice

Chapter 13
Ratio

Finding Ratios
Equivalent Ratios
Finding a Quantity
Comparing Three Quantities
Word Problems
Practice

Chapter 14
Rate

Unit Rate
Finding the Total Amount
    Given the Rate
Finding the Number of Units
    Given the Rate
Word Problems
Practice

Chapter 15
Percentage

Meaning of Percentage

Writing Percentages as
    Fractions in Simplest Form
Writing Decimals as
    Percentages
Writing Fractions as
    Percentages
Practice A
Percentage of a Quantity
Word Problems
Practice B
Review 3

Suggested number of class periods: 5 – 6

Lesson		Page	Resources	Objectives
	Chapter Opener	p. 5	TB: p. 1	Review the importance of one-to-one correspondence when counting.
1	Numbers to 10	p. 6	TB: p. 2 WB: p. 1	Count groups of objects from 1 to 10. Read numerals and number words to 10. Represent numbers to 10 on a ten-frame.
2	The Number 0	p. 9	TB: p. 6 WB: p. 3	Understand the concept of 0 as a set with 0 objects in it. Count back from 10 to 0.
3	Order Numbers	p. 11	TB: p. 9 WB: p. 7	Understand the sequence of numbers from 0 to 10. Order numbers to 10 from least to greatest and greatest to least.
4	Compare Numbers	p. 13	TB: p. 11 WB: p. 9	Compare numbers to 10.
5	Practice	p. 16	TB: p. 15 WB: p. 11	Practice ordering and comparing numbers to 10.
	Workbook Solutions	p. 19		

In **Dimensions Math® Kindergarten A**, students learned to:

- Count, read, and write whole numbers to 10.
- Compare groups of objects up to sets of 10 using comparison language: "greater/more than," "less/fewer than," or "the same as."
- Order and compare numbers to 10.

This chapter is both a review and an extension of content covered in the **Dimensions Math® Kindergarten** books. Most students can recognize, read, and write numbers, and count beyond 10. Chapter 1 is designed to reinforce basic number skills and familiarize students with materials. Students who have mastered that content may progress quickly through the first chapter, and may be ready to cover two lessons in one session.

The representation of an absence of a quantity as the quantity 0 is an important mathematical abstraction. However, as most people do not begin counting from 0, **Lesson 1: Numbers to 10** counts from 1 to 10, and 0 is taught is its own lesson.

Our written system for numeration is a base ten, or decimal, system. In a base ten system, there are 10 digits, 0 to 9, and their place in a number is based on powers of 10, e.g., ones, tens, hundreds, thousands, etc. So, the number that is one more than nine is written as 10. The 1 in the tens place denotes that it is 1 ten instead of 1 one. Thus, 35 means there are 3 tens and 5 ones.

This system allows for easy computations of large numbers by focusing on only single-digit operations.

Because this system is the basis for all standard algorithms in arithmetic, students are introduced to the concept from the beginning of their math education.

"Number" is used to describe a quantity. "Numeral" refers to a symbol from the set {0, 1, 2, 3, 4, 5, 6, 7, 8, 9}. Students may use the term "number" for both a number and the numeral at this level.

Comparison terms

"Fewer" as in "fewer than" is used when referring to a group of objects that we typically count (pets, toys, cars, etc.).

"Less" or "less than" is used to refer to aggregate quantity, such as "less milk" or "less sand."

"More than" is used in both cases.

"Greater than" or "less than" are used with numbers themselves.

Examples:
7 is greater than 3.
5 is less than 10.
I have fewer cats than dogs.
There is more sand than gravel.

Avoid using the terms "bigger" and "smaller" with students as that can lead to later confusion. For example, which is bigger:

3 or 7?

Model using "greater than" and "less than" with students, but do not require them to use proper wording in this chapter.

Materials

- Counters
- Playing cards
- Paper plates
- Pebble or other hopscotch marker
- Sidewalk chalk
- Painter's tape
- Linking cubes
- Stickers or bingo daubers
- Whiteboards

Note: Materials for Activities will be listed in detail in each lesson.

Blackline Masters

- Blank Ten-frame
- Ten-frame Cards
- Number Cards
- Color It, Trace It, Write It
- Make a Match Cards

Storybooks

- *Ten Black Dots* by Donald Crews
- *Fish Eyes: A Book You Can Count On* by Lois Ehlert
- *Zero Is the Leaves on the Tree* by Betsy Franco
- *My Little Sister Ate One Hare* by Bill Grossman
- *Count the Ways, Little Brown Bear* by Jonathan London
- *Mouse Count* by Ellen Stoll Walsh
- *Museum 123* by The New York Metropolitan Museum Of Art

Letters Home

- Chapter 1 Letter

Activities

Games and activities included in this chapter are designed to provide practice and extensions of counting and comparing numbers to 10. They can be used after students complete the Do questions, or anytime review and practice are needed.

Notes

Objective

- Review the importance of one-to-one correspondence when counting.

This **Chapter Opener** may be used as a full math lesson by reading a story from the suggestions at the beginning of the chapter or by completing the activities included.

Have students look at textbook page 1 and ask, "How many of each animal are there?"

Guide students to count out loud if they do not immediately know the quantities. If students have mastered counting to 10 accurately, continue straight to **Lesson 1: Numbers to 10**.

Whole class activities can be used as practice or review.

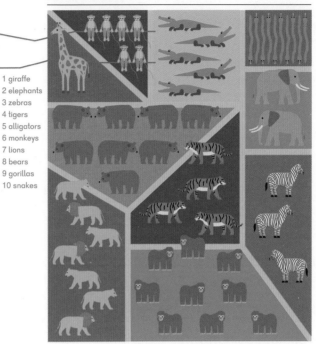

Chapter 1

Numbers to 10

1 giraffe
2 elephants
3 zebras
4 tigers
5 alligators
6 monkeys
7 lions
8 bears
9 gorillas
10 snakes

1

Activities

▲ Count the Classroom

Provide students with a list (in pictures or words) of objects in the classroom of which there are 10 or fewer. Have students find and count the items. For example:

_____ tables

_____ flags

_____ maps

_____ windows

▲ Scavenger Hunt

Have students find objects to match the given quantities. For example, ask them to find 3 yellow leaves. Collect different types of leaves, sticks, seeds, or other natural objects.

Have students write or draw pictures of quantities of objects in the classroom.

Lesson 1 Numbers to 10

Objectives

- Count groups of objects from 1 to 10.
- Read numerals and number words to 10.
- Represent numbers to 10 on a ten-frame.

Lesson Materials

- Counters, 10 per student
- Blank Ten-frames (BLM)

Think

Provide students with a small handful of counters and ask them to count them. Note if any students touch the same counters more than once and arrive at an incorrect answer. Ask students if there is any way to organize the counters so that they are easier to count.

Refer back to the **Chapter Opener** picture and ask students if there's an easier way to keep track of all the animals.

Provide each student with a Blank Ten-frame (BLM) for organizing their counters.

Have students count the number of each animal, starting with the giraffe. While counting, have students put a corresponding counter in a square on their ten-frame.

Have them remove all the counters from their ten-frame and repeat for the other rows.

Ask students what they notice about the patterns of filling in the ten-frame. Can they see 5 and some more?

To extend, ask students, "Are there other ways to make the numbers on a ten-frame? Is it easier to count when a row of 5 is already filled?"

Learn

In the textbook, students are relating the quantity to the representation in a ten-frame.

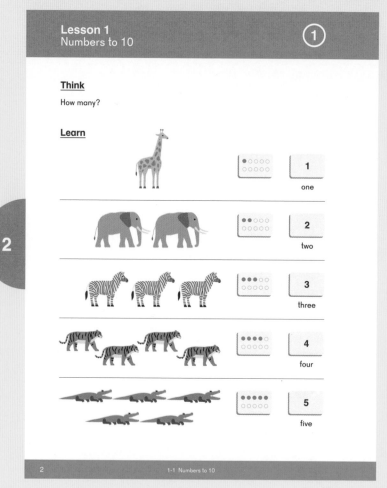

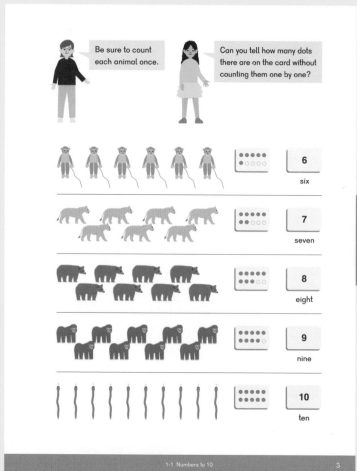

Do

1 Students that struggle may put counters on a Blank Ten-frame (BLM) to count. The page could also be shown on a screen and teachers can cross off each item as it is counted.

2 Students should begin to recognize some patterns. If one row of a ten-frame is complete, they might know it represents 5 objects without counting. Have students explain their thinking as they answer these questions.

For students who struggle with these tasks:

- Have them touch each dot on the picture of the ten-frame with their finger if needed.
- Provide counters and a Blank Ten-frame (BLM) to recreate the problems.

Activities

▲ Magic Thumb

Pointing your thumb up or down, have students chorally count up and down within 10 by ones.

For example, start out by saying, "Let's count by ones starting at 4. First number?" The class would respond, "4." Then, point your thumb up, and the class responds, "5." Then point your thumb down, and the class responds, "4." Point down again, and the class responds, "3," and so on.

▲ Count the Rockets

Play outside or in a gym. Have students make a large circle, then close their eyes. Select some students to run around inside the circle (with their eyes open) like rockets blasting off.

Students forming the circle open their eyes and try to count the students running in the middle of the circle.

This is a more difficult task than touching and counting.

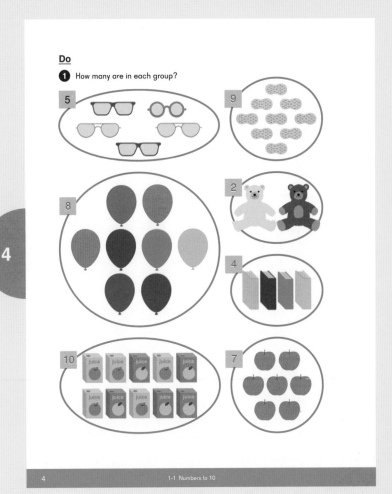

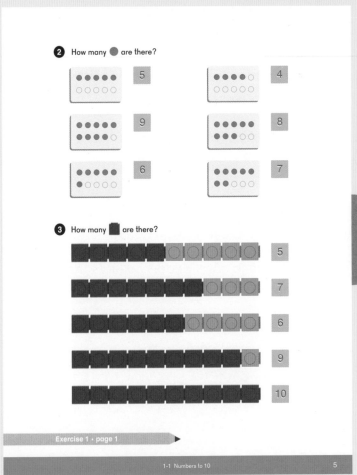

▲ Color It, Trace It, Write It

Materials: Color It, Trace It, Write It (BLM)

Give each student a copy of Color It, Trace It, Write It (BLM). Students color in the ten-frame, trace the number with a pencil, then write the numeral and number word.

★ Next Number Snap

Materials: A deck of playing cards with face cards removed, or Number Cards (BLM) 1 to 10, 2 or more sets

Play with up to four players. Deal out all Number Cards (BLM) to players facedown.

Players take turns turning over their top card and saying the number aloud. They put that card into a discard pile.

If the new card is one more than the top card on the discard pile, players say, "Snap."

The first person to say, "Snap," collects the discard pile.

The game ends when a player is out of cards. The player with the most cards wins.

Exercise 1 • page 1 ▶

Teacher's Guide 1A Chapter 1

Lesson 2 The Number 0

Objectives

- Understand the concept of 0 as a set with 0 objects in it.
- Count back from 10 to 0.

Lesson Materials

- Plate
- Counters
- Ten-frame Card (BLM) 0
- Number Word Card (BLM) 0
- Number Card (BLM) 0

Think

Show students a plate with five counters on it and ask them how many counters are on the plate.

Remove one counter and ask how many there are now. Continue to remove one counter at a time until there are no counters on the plate. Ask, "What number can we use to show there is nothing left on the plate?"

Show students the Number Word Card (BLM), Ten-frame Card (BLM), and Number Cards (BLM) for 0.

Learn

In the textbook, students are relating the quantity of none to the number, number word, and ten-frame representation.

Activities

▲ Hopscotch

Materials: Pebble or other marker for each player, sidewalk chalk or paper plates and painter's tape

Play outside or in a gym. Draw a hopscotch board using chalk, or tape down paper plates to create a hopscotch board with 0 as the starting spot.

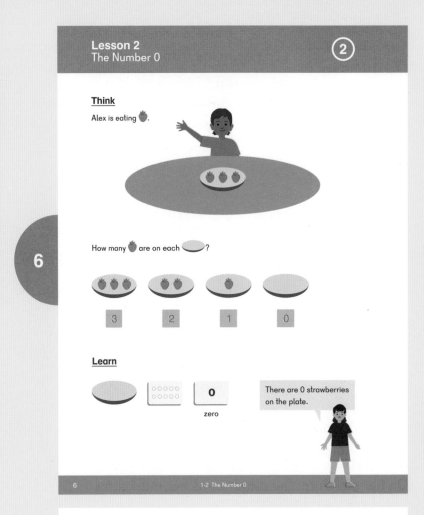

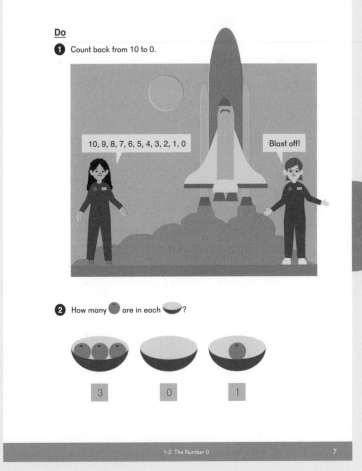

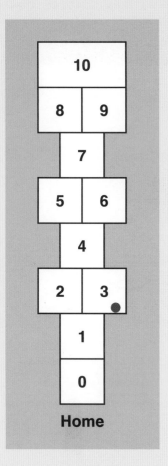

Players take turns standing in the 0 square (Home) square and tossing the marker. On their first turn, players aim for the 1 square. On each turn, players hop over the square with the marker and continue hopping in order, saying the numbers in each square aloud.

Square 10 is a rest stop. Players can put both feet down before turning around and hopping back to 0. Players pause in square 2 to pick up the marker from square 1, hop in square 1, and then out.

On her next turn, the player aims her marker for square 2, etc.

A player's turn is over if:

- Her marker does not land in the correct square.
- She loses her balance and puts a second foot down.
- She lands on a square where a marker is.

The winner is the first player to get through all 10 turns.

▲ Knock Number

This is a whole class activity that is slightly more difficult for students as they will need to hold counts in their head with no visual. A bell, chime, etc. can be used instead of knocking.

Have students put their heads down on their desk or table. The teacher knocks a certain number of times.

Students hold up fingers to show how many knocks.

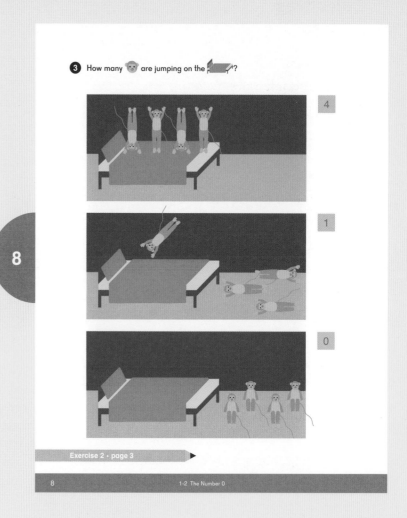

Exercise 2 · page 3

8   1-2 The Number 0

▲ Make a Match

Materials: Make a Match Cards (BLM), 0 to 10 only

This can be a whole class activity that uses number word, numeral, and ten-frame representations.

Pass out Make a Match Cards (BLM) to students and have them find cards that match their own card without speaking.

Students can also line up in order from 0 to 10.

For example, these cards would line up together:

Exercise 2 · page 3

Lesson 3 Order Numbers

Objectives

- Understand the sequence of numbers from 0 to 10.
- Order numbers to 10 from least to greatest and greatest to least.

Lesson Materials

- Linking cubes, 55 per student or pair of students
- Number Cards (BLM) 0 to 10, 1 set per student
- Blank Ten-frames (BLM) or index cards, 11 per student (to create their own ten-frame cards)
- Optional: stickers or bingo daubers

Think

Provide pairs of students with linking cubes and a set of Number Cards (BLM) from 0 to 10. Have students link together cubes to match the numbers on the cards, then ask them to organize their numbers.

Discuss the way students have organized their numbers and cubes.

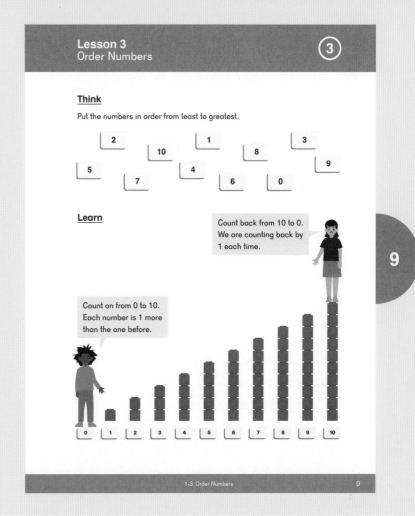

Learn

Ask students what they notice about how Dion and Mei organized their cubes.

Potential student answers:

- The numbers go up (increase) or down (decrease) by one.
- The blocks are different colors after the first 5.

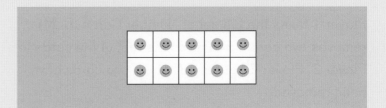

Using Blank Ten-frames (BLM) or index cards, have students create a set of ten-frame cards for numbers 0 to 10 by coloring in the correct number on each card. Alternately, students could use stickers or a bingo dauber to create a set of ten-frame cards.

Do

❶ — ❷ Students should use their ten-frame sets created in **Learn** to complete these questions.

❸ Have students call out or complete each row on whiteboards. They can recreate each problem with their Number Cards (BLM).

Activities

● Match

Materials: Number Cards (BLM) 0 to 10, Ten-frame Cards (BLM) 0 to 10

Lay cards in a faceup array. Have students match each Number Card (BLM) and Ten-frame Card (BLM) in order from greatest to least.

▲ Memory

Materials: Number Cards (BLM) 0 to 10, Ten-frame Cards (BLM) 0 to 10

Play using the same rules as **Match**, but set the cards out facedown in an array.

▲ What's Missing?

Materials: Number Cards (BLM) 0 to 10

Player 1 takes the full set of Number Cards (BLM), removes two cards, and gives the rest of the cards to Player 2. Player 2 figures out which two cards have been removed.

Exercise 3 • page 7 ▶

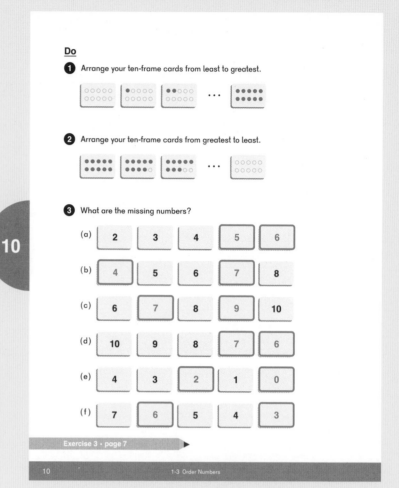

Lesson 4 Compare Numbers

Objective

- Compare numbers to 10.

Lesson Materials

- Two-color counters, 20 per each student

Think

Have pairs of students work together. They should use counters to represent the dogs and cats in the **Think** problem. Discuss how students can decide if there are more cats or more dogs.

Learn

Have students use one color counter to represent the dogs and the other color to represent the cats. Have them line the counters up next to each other similar to the **Learn** example, so that they can easily see the one-to-one correspondence: one cat to one dog, one cat to one dog. Students will see that there is one **more** dog than cat.

Introduce the term "**fewer**." Students may be more familiar with the term "less." Encourage them to answer with both statements:

- There are more ___ than ___.
- There are fewer ___ than ___.

Give students other quantities of dogs and cats to practice comparing. For example, "What if there were 3 dogs and 1 cat? What if there were 3 cats and 0 dogs?"

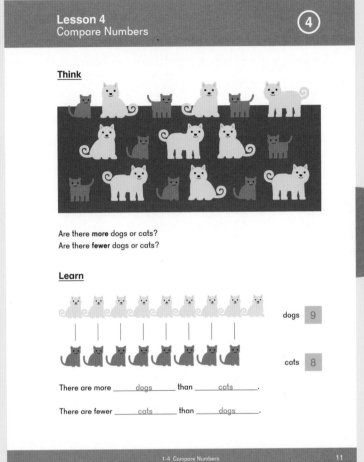

Do

1 (a) and **2** (a) These questions assess understanding of "more" and "fewer" with one-to-one correspondence and a direct comparison.

1 (b) and **2** (b) These questions assess a deeper understanding without the direct comparison, while still using a picture. For students struggling with the (b) problems, have them note what is the same in the fruit arrangement, namely the pattern of five fruits. Then have them concentrate on the remaining differences.

3 Students should write "more" or "fewer."

4 — **6** These questions on page 14 add in the terms "greater" and "less."

Students who struggle can use two different colored counters matched to the illustrated items to determine which group has more items.

Activities

● **More Face-off or Fewer Face-off**

Materials: Ten-frame Cards (BLM) 0 to 10, Optional: linking cubes or counters, ▲ Number Cards (BLM) 0 to 10

Players each flip over a card at the same time. The greatest Ten-frame Card (BLM) wins (or least, depending on version of game).

Students can use linking cubes or counters to see whose number is greater.

▲ Play with Number Cards (BLM).

★ Have the winner say how many greater or less his card is than the other player's card.

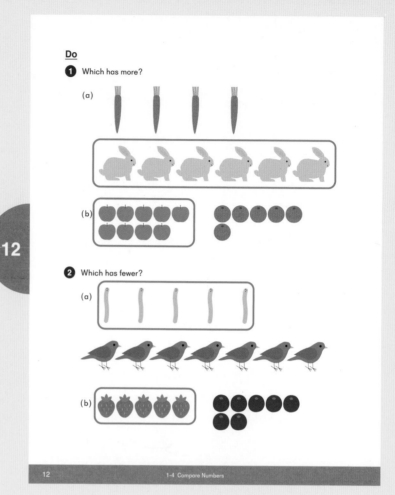

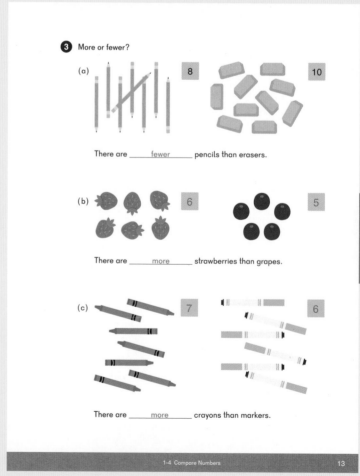

▲ Rock, Paper, Scissors, Math!

This game can be taught at any point during the lesson. Different versions will be played throughout the coming chapters.

Encourage students to play during down time in class or at recess, or use it as a quick assessment by playing with a student. The game can be played with two students (or three for added difficulty).

Players say, "Rock, paper, scissors, math!"

On the word "math," each player shoots out some fingers. The student who says the greater number of fingers first wins.

For example, if Player 1 shows 6 fingers and Player 2 shows 9 fingers, the first player to say, "9," is the winner.

Students can also play to the lesser number. In the previous example, the first player to say, "6," wins.

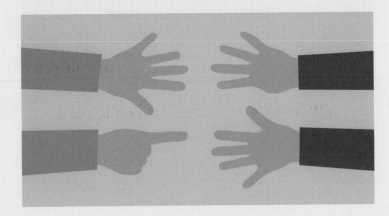

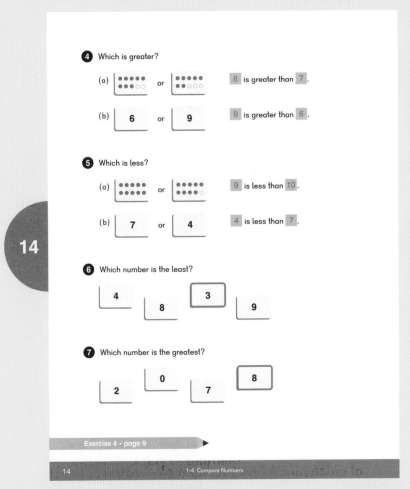

★ Greater Than or Less Than?

Materials: Set of Ten-frame Cards (BLM) 0 to 10, Number Cards (BLM) 0 to 10, or playing cards for each player

Player 1 selects a secret card from her hand and places it facedown.

Player 2 tries to guess the number on the secret card by laying down a card from his hand faceup.

Player 1 then tells whether the secret card is greater or less than Player 2's faceup card.

Player 2 continues selecting and showing different cards until he finds the value of the secret card. Players switch roles.

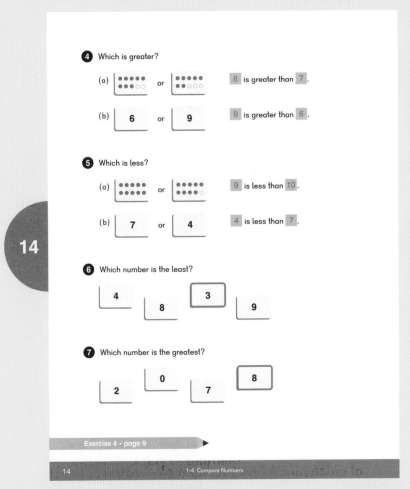

Exercise 4 · page 9

Objective

- Practice ordering and comparing numbers to 10.

Practice

After students complete the **Practice** in the textbook, have them continue to work with numbers to 10 by playing games from this chapter.

Activities

● **Number Line Walk**

Materials: Sidewalk chalk or painter's tape and paper plates

Create a large number path outside with chalk or inside with painter's tape. You could also tape numbered plates on the floor.

Have students step on a number that is greater or less than a given number. You may have several students on a number.

▲ Create several number lines and have students stand one to a number. If a student's first choice is filled, she must find another number that works.

★ Have students stand on a number that is 1 more than, 2 more than, 1 less than, 2 less than, etc. a given number.

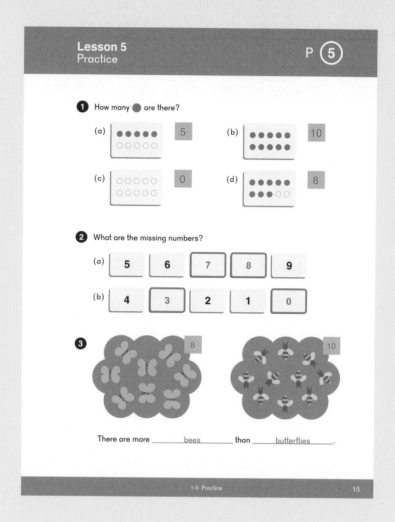

▲ Ruckus

Materials: Deck of cards

Deal each player seven cards. Players immediately place all cards of the same value faceup (for example, two 4s or three Jacks).

If another person has a card of that value, she can put it down on the pile and take the pile to her part of the table. All players do this at the same time.

Once all play has stopped, the dealer hands out new cards, and the pile building and taking is repeated until all cards have been dealt.

The player with the most cards at the end wins.

★ Rummy

Materials: Deck of cards

Deal each player 7 cards. The remaining cards go in the middle in the draw pile. The dealer turns the top card faceup and sets it next to the draw pile to start a discard pile.

Players take turns drawing from either the draw or discard pile to collect sets or runs of three or four cards in their hands. Each turn consists of a draw and a discard.

A run consists of consecutive numbers in the same suit, for example, the 3, 4, and 5 of spades. A set is made up of three or four cards bearing the same number, regardless of suit (three 4s). When a student collects two runs, two sets, or one of each, she lays down her cards and is the winner.

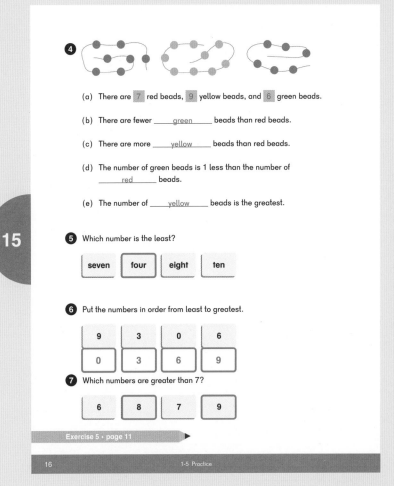

15

4

(a) There are 7 red beads, 9 yellow beads, and 6 green beads.

(b) There are fewer ____green____ beads than red beads.

(c) There are more ____yellow____ beads than red beads.

(d) The number of green beads is 1 less than the number of ____red____ beads.

(e) The number of ____yellow____ beads is the greatest.

5 Which number is the least?

| seven | four | eight | ten |

6 Put the numbers in order from least to greatest.

| 9 | 3 | 0 | 6 |
| 0 | 3 | 6 | 9 |

7 Which numbers are greater than 7?

| 6 | 8 | 7 | 9 |

Exercise 5 · page 11

16   1-5 Practice

Exercise 5 · page 11

Notes

Chapter 1 Numbers to 10

Exercise 1

Basics

1 Match.

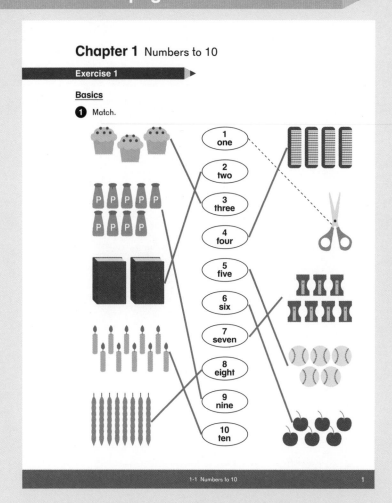

Practice

2 Color how many.

Check that students have colored the correct number.

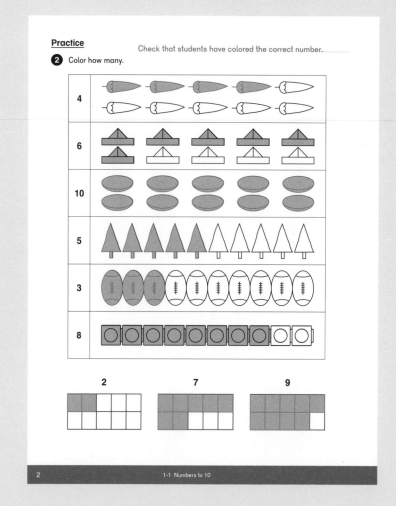

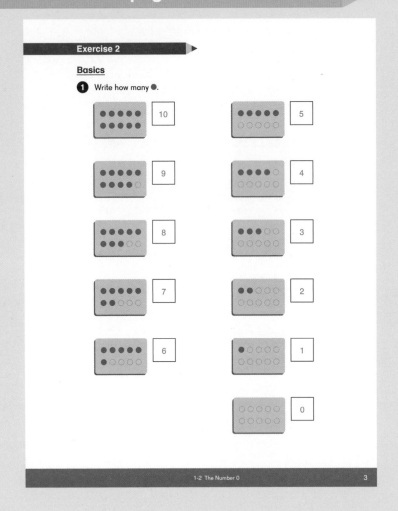

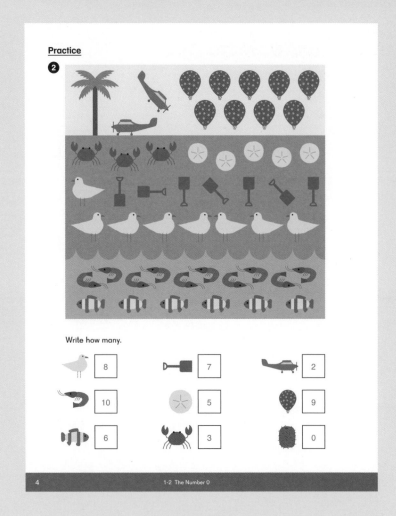

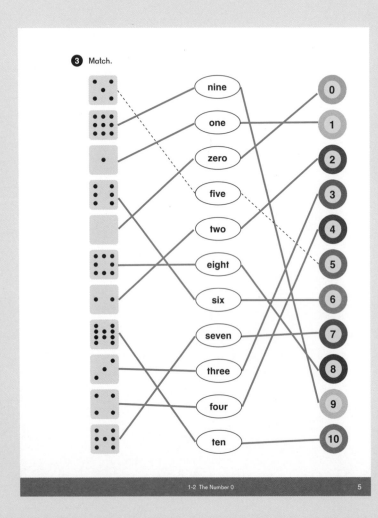

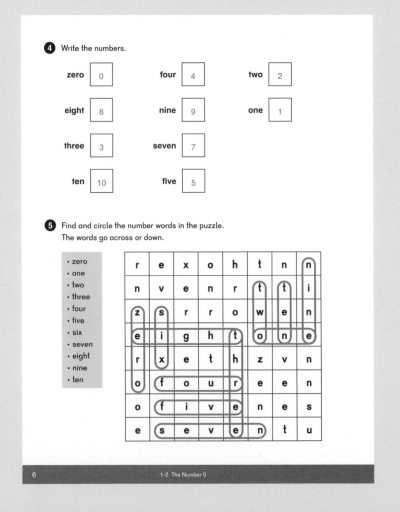

Exercise 3

Basics

1 Write the missing numbers.

0	1	2	**3**	4	5	6	7	8	**9**	10

Practice

2 Count back.
Write the numbers.

10	9	8	7	6	5	4	3	2	1	0

3 Write the missing numbers.

(a) | **3** | 4 | **5** | 6 | **7** | 8 |

(b) | **7** | **6** | 5 | 4 | **3** | 2 |

4 Write the number that is...

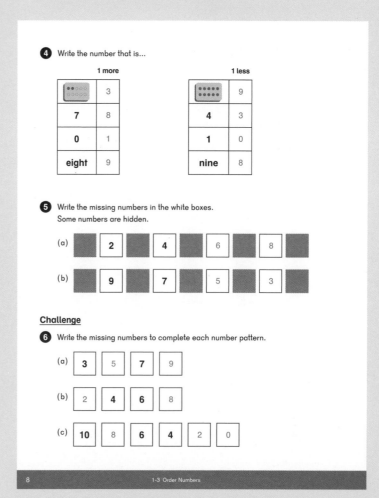

	1 more
●●●●● / ●●●●●	3
7	8
0	1
eight	9

	1 less
●●●●●	9
4	3
1	0
nine	8

5 Write the missing numbers in the white boxes.
Some numbers are hidden.

(a) | | **2** | | **4** | | 6 | | 8 | |

(b) | | **9** | | **7** | | 5 | | 3 | |

Challenge

6 Write the missing numbers to complete each number pattern.

(a) | **3** | 5 | **7** | 9 |

(b) | 2 | **4** | **6** | 8 |

(c) | **10** | 8 | **6** | **4** | 2 | 0 |

Exercise 4

Basics

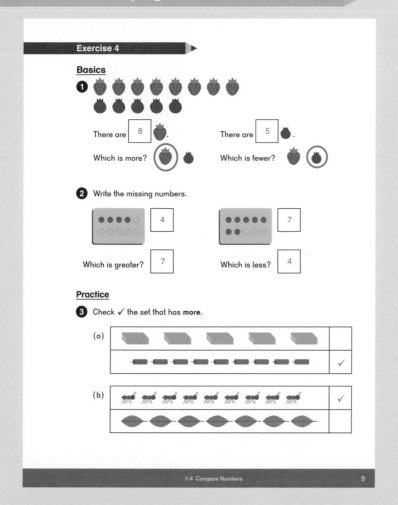

1

There are | 8 | 🍓 .   There are | 5 | 🫐 .

Which is more? 🍓   Which is fewer? 🫐

2 Write the missing numbers.

| 4 | Which is greater? | 7 |

| 7 | Which is less? | 4 |

Practice

3 Check ✓ the set that has **more**.

(a) (row of 5 shapes) | |
(row of 9 small shapes) | ✓ |

(b) (row of 8 ants) | ✓ |
(row of 7 shapes) | |

4

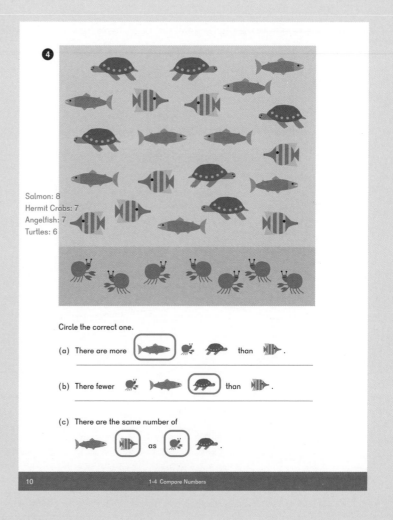

Salmon: 8
Hermit Crabs: 7
Angelfish: 7
Turtles: 6

Circle the correct one.

(a) There are more (🐟 fish) than 🐠 .

(b) There fewer 🦀 🐟 (🐢) than 🐠 .

(c) There are the same number of
(🐟) (🐠) as 🦀 🐢 .

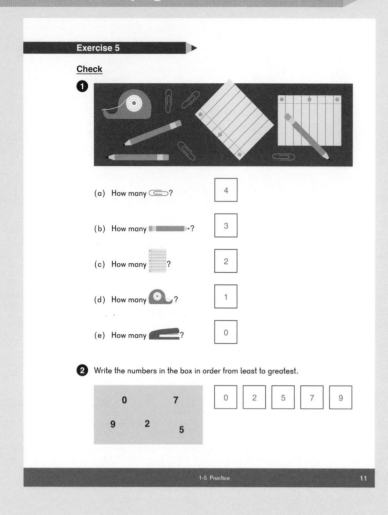

Exercise 5

Check

1

(a) How many ⌐◯? 4

(b) How many ▰▬▬▶? 3

(c) How many ▤? 2

(d) How many ◉? 1

(e) How many ◢? 0

2 Write the numbers in the box in order from least to greatest.

0	7
9	2
	5

0 2 5 7 9

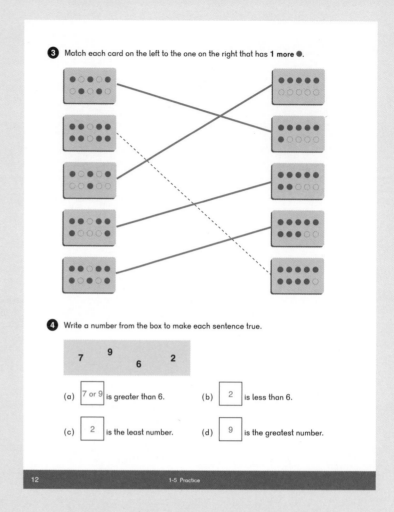

3 Match each card on the left to the one on the right that has **1 more** ●.

4 Write a number from the box to make each sentence true.

7	9		
		6	2

(a) 7 or 9 is greater than 6.   (b) 2 is less than 6.

(c) 2 is the least number.   (d) 9 is the greatest number.

Students can write the numbers of each type next to one of the objects of that type.

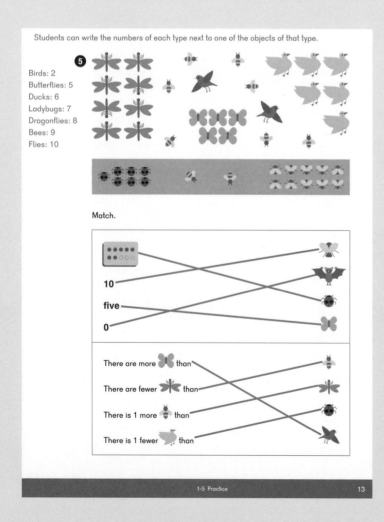

5

Birds: 2
Butterflies: 5
Ducks: 6
Ladybugs: 7
Dragonflies: 8
Bees: 9
Flies: 10

Match.

10
five
0

There are more 🦋 than
There are fewer 🌿 than
There is 1 more 🐝 than
There is 1 fewer 🐦 than

Challenge

6 Cross out the greatest number.
Circle all the numbers that are less than 6.

③ ⑤ ✗ ④ 7 8 ⓪ ②

7 What number is...

(a) 2 less than 7? 5

(b) 2 more than 7? 9

(c) 2 less than 2? 0

(d) 3 more than 0? 3

8 (a) What numbers are greater than 4 and less than 9?

5, 6, 7, 8

(b) What numbers are less than 6 and greater than 3?

4, 5

Suggested number of class periods: 7 – 8

Lesson	Page	Resources	Objectives
Chapter Opener	p. 27	TB: p. 17	Review number bonds for numbers to 5.
1 Make 6	p. 28	TB: p. 18 WB: p. 15	Associate number bonds with composing and decomposing whole numbers. Find number pairs that make 6 and represent them with number bonds. Find the missing part of a number bond.
2 Make 7	p. 30	TB: p. 20 WB: p. 17	Find number pairs that make 7 and represent them with number bonds.
3 Make 8	p. 32	TB: p. 22 WB: p. 19	Find number pairs that make 8 and represent them with number bonds.
4 Make 9	p. 34	TB: p. 24 WB: p. 21	Find number pairs that make 9 and represent them with number bonds.
5 Make 10 — Part 1	p. 36	TB: p. 26 WB: p. 23	Find number pairs that make 10 and represent them with number bonds.
6 Make 10 — Part 2	p. 38	TB: p. 28 WB: p. 25	Find number pairs that make 10 and represent them with number bonds.
7 Practice	p. 41	TB: p. 31 WB: p. 27	Practice working with number bonds to 10.
Workbook Solutions	p. 43		

This chapter builds on content covered in **Dimensions Math® Kindergarten B, Number Bonds**. Lessons in this chapter reinforce basic number skills and familiarize students with number bonds to 10. These lessons are of particular importance as students will be using number bonds to show the strategies they use to add and subtract two-digit numbers in future lessons.

Key Points

Having students create number stories from lessons lays the foundation for working with word (story) problems later in the materials.

Parts go together to make the whole. It doesn't matter if the parts are on the left or right, top or bottom, or what shape is used to set off the numbers from each other. These lessons will show number bonds in a variety of orientations to help students learn that the relationship between the number, not their visual presentation, is important.

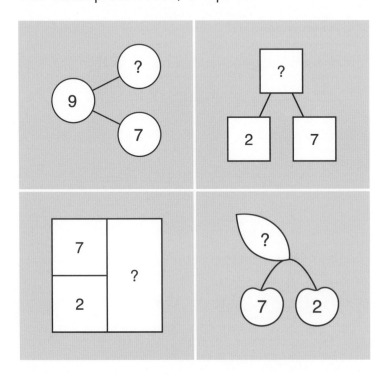

The order of the parts in a number bond does not matter. Both, "5 is 2 and 3," and, "5 is 3 and 2," are acceptable and reinforce the commutative property of addition.

Number bonds generally refer to a whole and only two parts. Students may be curious about drawing more parts. For example, "3 and 1 and 2 make 6" would be represented by three parts attached to the whole. Items of the same type or kind can be associated using number bonds. We refer to these types as "units." 4 cats and 3 dogs can be parts when the whole is pets; boys and girls are parts when the whole is children.

This chapter uses "____ and ____ make____" instead of the formal addition and equal symbols (+ and =) which will be introduced in **Chapter 3**: **Addition**. Students may recall the symbols from **Dimensions Math® Kindergarten** and should not be discouraged from using them.

This series will use circles around the numbers in the number bond initially to set apart each number, but later the circles will be dropped, particularly around the whole.

By the end of **Chapter 4**: **Subtraction**, students should know their number bonds to 10 to automaticity, that is, when given two parts, the students know without counting what the whole is. When given one part and a whole, a student can tell the missing part. There are many practice games and activities to help students achieve this goal. The games and activities can be used continuously with students who have not mastered their number bonds to 10.

Materials

- Two-color counters
- Cups
- 3 large hula hoops
- Sidewalk chalk
- Painter's tape
- Paper plates
- Colored pencils or crayons
- Dice, 6-sided or 10-sided
- Playing cards
- Linking cubes
- Dried beans
- Blocks
- Dominoes
- Whiteboards

Note: Materials for Activities will be listed in detail in each lesson.

Blackline Masters

- Number Bond Story Template
- Number Cards
- Ten-frame Cards
- Number Bonds to 10 Flash Cards

Letters Home

- Chapter 2 Letter

Notes

Objective

- Review number bonds for numbers to 5.

Lesson Materials

- Two-color counters, 5 per student
- Number Bond Story Template (BLM), 1 per student

Chapter 2

Number Bonds

17

Have students look at the **Chapter Opener**. Point out the number bonds and ask students what they might represent about the polar bears.

Provide students with 5 two-color counters and a Number Bond Story Template (BLM), and have them model what they see with the polar bears.

Students should observe that:

- 3 are in the water, 2 are out of the water.
- 4 have their mouths closed, 1 has its mouth open.
- 2 are adults, 3 are cubs.
- 1 has his pink tongue out and 4 do not.

Discuss all four situations. "**Part** of the polar bears are in the water. How many are in the water? **Part** of the polar bears are out of the water. How many are out of the water? How many polar bears are there **altogether**?"

When writing the number bonds on the board, the parts can be labeled:

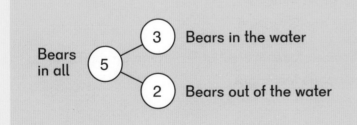

Bears in all — 5 — 3 Bears in the water — 2 Bears out of the water

Be sure to point out a number bond where a part is 0. Prompt students to tell a story involving 5 bears and 0 bears as parts: "5 white bears and 0 brown bears."

If necessary, extend this lesson with the activity **Make Number Stories** for practice, or continue to **Lesson 1: Make 6**.

Activity

▲ Make Number Stories

Materials: Number Bond Story Template (BLM) or art paper, markers or crayons

Assign or have students select a number bond with a whole of 5. Have them write the number bond, then illustrate a number story for their number bond. They do not need to write out the story.

For example, a student might draw 1 brown cat and 4 white cats.

Students can use the Number Bond Story Template (BLM), or be creative in drawing their own bonds.

Students can share their number stories.

Objectives

- Associate number bonds with composing and decomposing whole numbers.
- Find number pairs that make 6 and represent them with number bonds.
- Find the missing part of a number bond.

Lesson Materials

For each student:

- Blank Number Bond (BLM)
- 6 two-color counters
- Number Cards (BLM) 0 to 6
- Ten-frame Cards (BLM) 0 to 6

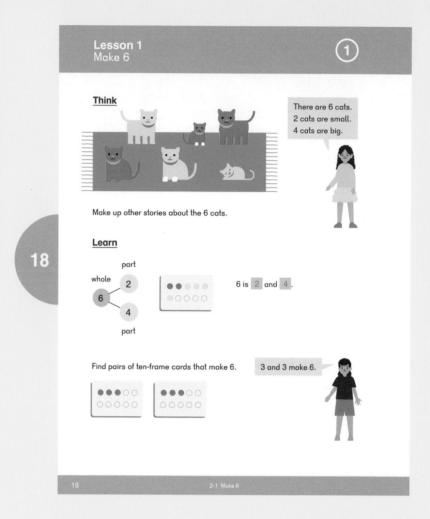

Think

Have students discuss Sofia's statement about the cats. Provide students with two-color counters and a Blank Number Bond (BLM). Have them model Sofia's statement about the cats by placing counters in the two parts to represent big cats and small cats.

Have students make up other number stories about the cats and model them with the counters on their number bonds using a different color for each part. Possible student responses include:

- There are 3 gray cats and 3 orange cats.
- There are 5 cats awake and 1 cat asleep.
- There are 6 cats with collars and 0 cats without collars.
- 3 cats have yellow collars and 3 cats have blue collars.

While students are sharing observations, various forms of phrasing are correct. For example:

- There are 6 cats. 2 cats are small. 4 cats are big.
- There are 6 cats. 4 cats are big. 2 cats are small.

Remind students that since the total is the same, the same number bond represents both situations.

Learn

Choose several student number stories and have students represent them with counters. Write the corresponding number bonds on the board:

- 1 and 5 make 6.
- 2 and 4 make 6.
- 3 and 3 make 6.

Provide students a set of Ten-frame Cards (BLM) to complete Mei's task.

Do

1 Have students use Ten-frame Cards (BLM) 0 to 6, or they can use those created in **Chapter 1**: **Lesson 3**.

Hand out Number Cards (BLM) 0 to 6.

2 Ensure students understand that the two parts go together to make the whole. It doesn't matter if the parts are on the left or right, top or bottom.

Go Fish for 6 and **Under the Cup** activities can be played with partners and used for the rest of the lessons in Chapter 2 by simply changing the quantity of cards or counters.

Activities

▲ Go Fish for 6

Materials: Number Cards (BLM) 0 to 6, 4 shuffled sets

Deal five cards each to up to 4 players. Players take turns asking for a number that, when paired with a card they're holding, makes 6.

When Player 1 has a pair of Number Cards (BLM) that make 6, he lays them down. If Player 1 asks for a card and the opponent asked does not have it, the opponent says, "Go Fish" and Player 1 draws one card from the draw pile. Play continues clockwise. The first player to pair all of their cards is the winner.

▲ Under the Cup

Materials: 6 counters, cup

Player 1 hides some counters under the cup and puts the rest on the table. Player 2 counts what is on the table and tells how many counters are hiding.

★ Play with 6 linking cubes assembled into a tower. Player 1 breaks the tower behind his back. He shows one part of the tower to Player 2. Player 2 says how many cubes are hidden behind Player 1's back.

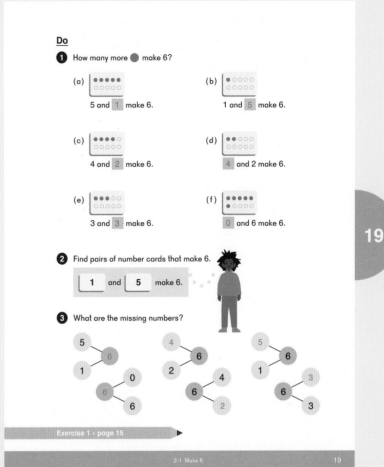

Exercise 1 • page 15

Lesson 2 Make 7

Objective

- Find number pairs that make 7 and represent them with number bonds.

Lesson Materials

For each student:

- Blank Number Bond (BLM)
- 7 two-color counters
- Ten-frame Cards (BLM) 0 to 7

Think

Bring 7 students to the front of the room and have students make number stories with 7. Find obvious differentiating qualities:

- Boots/sneakers
- Glasses/no glasses
- Boys/girls
- Short sleeves/long sleeves

As students create number stories, write the corresponding number bonds on the board.

Have students discuss Alex's statement about the apples. Provide students with two-color counters and a Blank Number Bond (BLM). Have them model Alex's statement about the apples.

Have students make up other number stories about the apples and model them with the counters on their number bonds using a different color for each part. Examples:

- 1 apple has a bite out of it. 6 apples are whole.
- 2 apples have worms. 5 apples have no worms.
- 4 apples have leaves. 3 apples have no leaves.

Prompt students to create a number story with a 0 part. For example, "There are 7 apples and 0 peaches. There are 7 fruits in all."

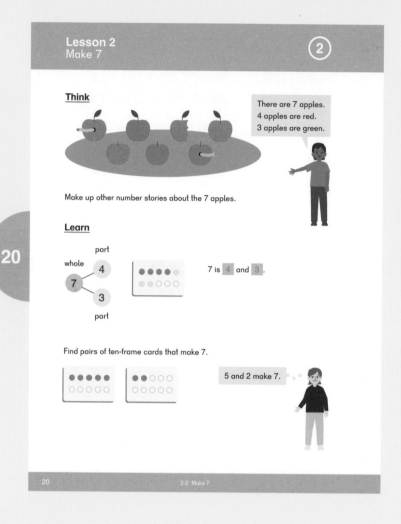

Learn

While students are sharing number stories, write the corresponding number bonds on the board.

Activities from prior lessons can be adapted for 0 to 7.

Provide students a set of Ten-frame Cards (BLM) to complete the **Learn** task.

Do

1 Have students use the Ten-frame Cards (BLM) for 0 to 7 if needed.

2 Ask students to picture how many more are needed to make 7 (or, "What is the whole?").

Activity

▲ Human Number Bonds

Materials: 3 large hula hoops (you may want to tape them down for safety as students will be creating human number bonds in them), painter's tape

Have students start in the two hula hoops representing parts, then have them step into the hula hoop representing the whole. Students watching should write the parts on a whiteboard.

If the hula hoops are too small, or unavailable, tape a giant number bond on the floor or use blocks or other classroom materials. Put a whiteboard with the number 7 in the whole and have students make different number bonds with the blocks or materials in the parts.

Exercise 2 • page 17

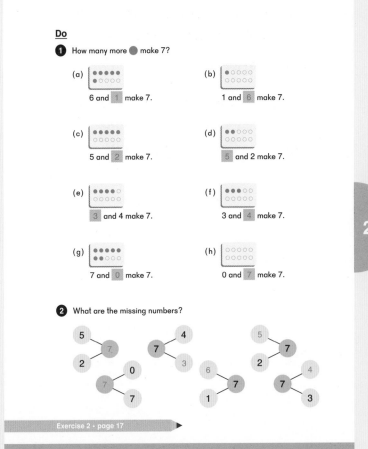

Lesson 3 Make 8

Objective

- Find number pairs that make 8 and represent them with number bonds.

Lesson Materials

- Number Bond Story Template (BLM)
- Two-color counters, 8 per student
- Ten-frame Cards (BLM) 0 to 8

Think

Have students discuss Dion's statement about the flowers. Provide students with two-color counters and a Number Bond Story Template (BLM). Have them model Dion's statement about the flowers.

Have students make up other number stories about the flowers and model them with the counters on their number bonds using a different color for each part. Examples:

- 1 flower has no leaves. 7 flowers have leaves.
- 2 flowers are small. 6 flowers are big.
- 4 flowers have bees on them. 4 flowers do not have bees.
- 4 flowers are on the green hill. 4 flowers are on the brown hill.

See if students can come up with a situation for 8 and 0.

Learn

While students are sharing number stories for 8, write the corresponding number bonds on the board. (This should be routine for them by this lesson.) Students can also review number bond combinations from prior lessons.

Activities from prior lessons can be adapted for use with number bonds to 8.

Provide students a set of Ten-frame Cards (BLM) to complete the **Learn** task.

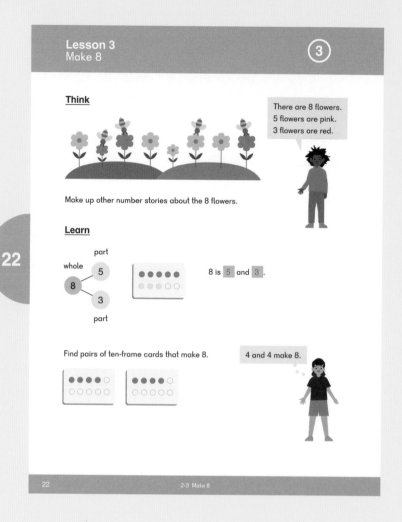

Do

1 Have students use Ten-frame Cards (BLM) for 0 to 8 if needed.

2 Ask students to picture how many more are needed to make 8 (or, "What is the whole?").

Activities

● Number Bond Match

Materials: Number Cards (BLM) and Ten-frame Cards (BLM) 0 to 8, 1 or 2 sets of each

Lay out cards faceup in an array. Have students choose a card and look for the complement to 8.

▲ Number Bond Memory

Materials: Number Cards (BLM) and Ten-frame Cards (BLM) 0 to 8, 1 or 2 sets of each

Play using the same rules as **Match**, but set the cards out facedown in an array. Playing this version will have students making many more matches as they must find a pair of cards that make 8.

▲ Go Fish for 8

Materials: Number Cards (BLM) 0 to 8, 4 shuffled sets

Deal 5 Number Cards (BLM) each to up to 4 players. Players take turns asking for a number that, when paired with a card they're holding, makes 8.

When Player 1 has a pair of cards that make 8, he lays them down. If Player 1 asks for a card and the opponent asked does not have it, the opponent says, "Go Fish" and Player 1 draws one card from the draw pile. Play continues clockwise. The first player to pair all of their cards is the winner.

Modify previous games from this chapter for 0 to 8:

- **Under the Cup**
- **Human Number Bonds**

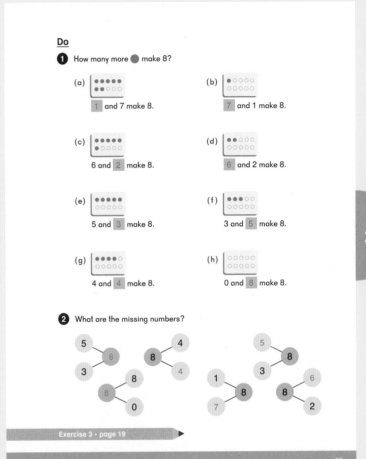

▲ Show Me the Bond

Use this whole class activity as a quick check for understanding.

With your fingers, show students part of a number bond to 8. Have students show you their part of the bond. For example, tell students, "We're going to make number bonds to 8. My part is (show five fingers). Show me, on your hands, what your part is."

Exercise 3 · page 19

Objective

- Find number pairs that make 9 and represent them with number bonds.

Lesson Materials

- Two-color counters, 9 per student
- Ten-frame Cards (BLM) 0 to 9
- Blank Number Bond (BLM)

Think

Have students discuss Sofia's comments about the ducks. Provide students with two-color counters and a Blank Number Bond (BLM), and have them model other number stories about the ducks. See if students can come up with a situation for 9 and 0.

Examples of stories that make 9:

- 3 ducks are quacking and 6 ducks are not quacking.
- 1 duck is brown. 8 ducks are gray (white).
- 2 ducks are flapping their wings. 7 ducks are not flapping their wings.

Record all number bonds on the board.

Learn

Have students find pairs of Ten-frame Cards (BLM) that make 9.

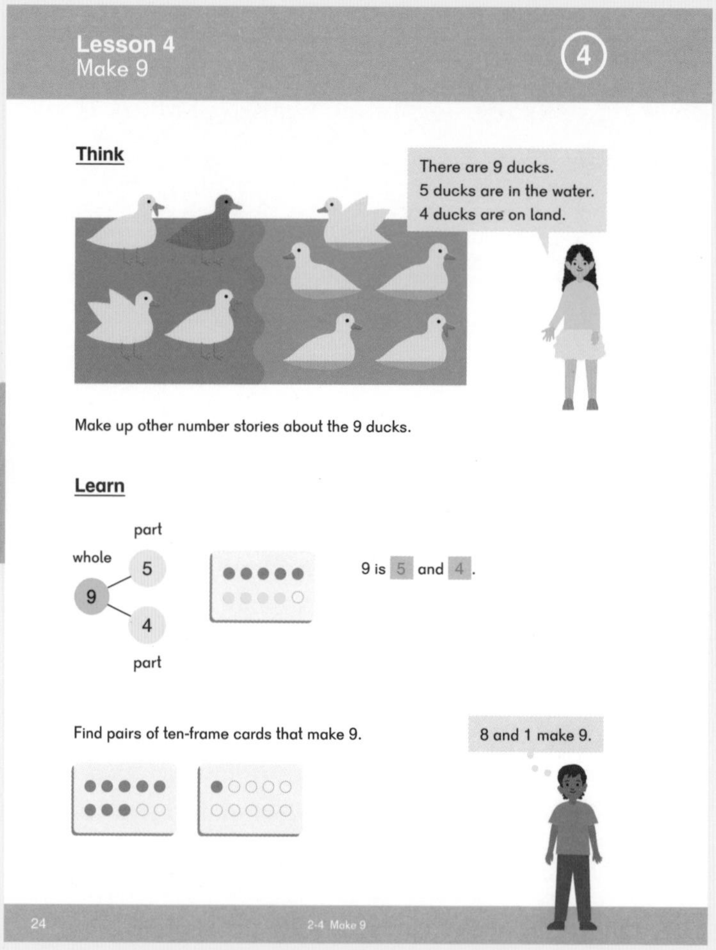

Lesson 4
Make 9
④

Think

There are 9 ducks.
5 ducks are in the water.
4 ducks are on land.

Make up other number stories about the 9 ducks.

Learn

whole — part 5, 9, 4 — part

9 is 5 and 4.

Find pairs of ten-frame cards that make 9.

8 and 1 make 9.

24   2-4 Make 9

Do

2 Ask students to picture how many more are needed to make 9 (or, "What is the whole?").

Activities

▲ Family Bonds

Materials: Art paper, markers or crayons

Creating **Family Bonds** is a great way to have students create math art to share on the wall or with their family.

Have students create **Family Bonds** with the number of adults and children in their household. The number of adults will be one part and the number of children will be the other part. They can draw pictures of family members in each part and the total number in the whole.

Teachers may need to work with students who have more than 9 members in their family.

▲ Number Bond Hop

Materials: Sidewalk chalk or paper plates and painter's tape

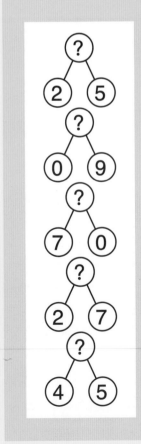

Create number bonds using chalk on a playground or sidewalk, or paper plates and painter's tape on a classroom floor. Have students say the parts and whole as they hop along the trail.

Use bonds that make 9 or all bonds learned.

★ For an added challenge, provide a whole and have a part missing.

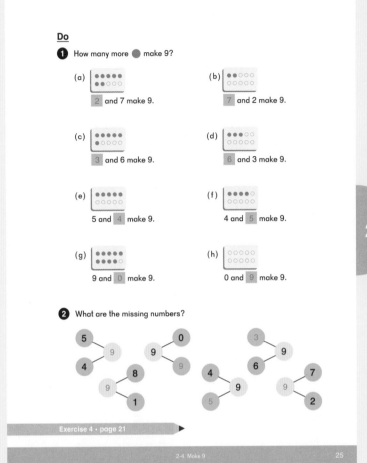

Do

1 How many more ● make 9?

(a) 2 and 7 make 9.

(b) 7 and 2 make 9.

(c) 3 and 6 make 9.

(d) 6 and 3 make 9.

(e) 5 and 4 make 9.

(f) 4 and 5 make 9.

(g) 9 and 0 make 9.

(h) 0 and 9 make 9.

2 What are the missing numbers?

Exercise 4 • page 21

2-4 Make 9   25

Modify previous games from this chapter for 0 to 9:

- **Go Fish**
- **Under the Cup**
- **Human Number Bonds**
- **Show Me the Bond**
- **Number Bond Match**
- **Number Bond Memory**

Exercise 4 • page 21

Lesson 5 Make 10 — Part 1

Objective

- Find number pairs that make 10 and represent them with number bonds.

Lesson Materials

- Ten-frame Cards (BLM) 0 to 10, 1 set per student
- Linking cubes or dried beans

Think

Have students discuss Emma's statement about the frogs and find other number stories similar to prior lessons. Record number bonds on the board.

Examples of stories that make 10:

- 5 frogs are eating flies and 5 frogs are not.
- There are 4 yellow frogs and 6 green frogs.
- 3 frogs are big and 7 frogs are small.
- 2 frogs have spots and 8 have no spots.
- 1 frog is jumping and 9 frogs are sitting.

Learn

Have students find pairs of Ten-frame Cards (BLM) that make 10.

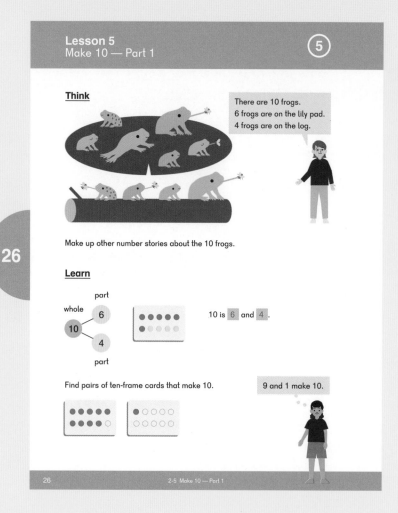

26

Do

1 Have students use 0 to 10 of the Ten-frame Cards (BLM) if needed.

2 Use beans, counters, or linking cubes to play the game. (See page 29 of this Teacher's Guide for **Under the Cup** directions from **Lesson 1: Make 6**.)

Activities

▲ Super Number Bonds

Materials: Sidewalk chalk or paper plates and painter's tape

To play outdoors, create multiple decompositions with chalk on sidewalk. To play indoors, create **Super Number Bonds** with paper plates and painter's tape.

Have students find different ways of completing the **Super Number Bonds**. This activity could also be used as a pencil and paper task, as seen in the workbook on page 26.

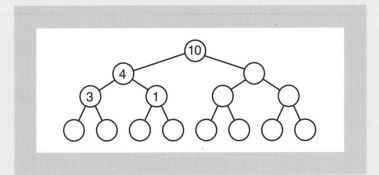

▲ Rock, Paper, Scissors, Math!

Partners play a different version of **Rock, Paper, Scissors, Math** that was originally taught in **Chapter 1: Lesson 4**.

On the word, "Math," each player shoots out some fingers on one hand. The student who says the sum of the fingers first is the winner.

For example, if Player 1 shows 4 fingers and Player 2 shows 4 fingers, the first player to say, "8," is the winner.

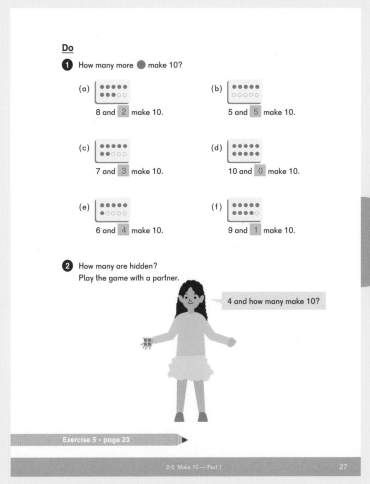

Modify previous games from this chapter for 0 to 10:

- **Go Fish**
- **Under the Cup**
- **Human Number Bonds**
- **Show Me the Bond**
- **Number Bond Match**
- **Number Bond Memory**
- **Number Bond Hop**

Exercise 5 · page 23

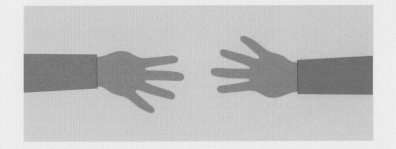

Lesson 6 Make 10 — Part 2

Objective

- Find number pairs that make 10 and represent them with number bonds.

Lesson Materials

For each pair of students:

- 20 linking cubes or blocks, 10 each of 2 colors
- Number Cards (BLM) 1 to 10
- Die, 6-sided or 10-sided

Think

Give students two colors of linking cubes and have them put them together to show ways to make 10.

Learn

Have students write number bonds for each of the combinations. Students should realize that the number bond for 1 blue and 9 orange, for example, is the same as the one for 9 blue and 1 orange.

Ask them to think of a way to make 10 that is not included on the page. Students may show all of one color of cubes for 10 and 0.

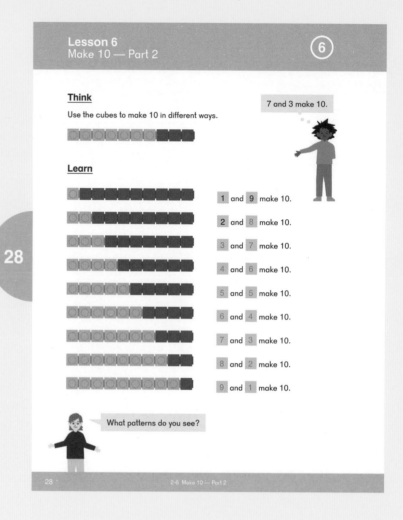

Teacher's Guide 1A Chapter 2   © 2017 Singapore Math Inc.

Do

1 This question can be used as a Small Group or Center Activity. Have students use two sets of Number Cards (BLM) 0 to 10, or work with a partner.

3 Students can roll a 6-sided or 10-sided die, or flip over number cards.

4 See directions from **Memory** game from **Chapter 2: Lesson 3** and modify for numbers to 10.

Activities

★ Towers

Materials: Blocks or linking cubes, 10 that are all the same size

This can be done as a whole group activity or an independent challenge. Stack 10 blocks to make any number of towers. The height of each tower must be different. Encourage students to find more than one solution.

For example, a student may make two towers, one of 4 blocks and the other of 6 blocks. He could also make three towers of 6 blocks, 3 blocks, and 1 block.

Have them record their solutions for discussion.

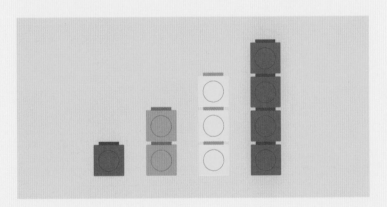

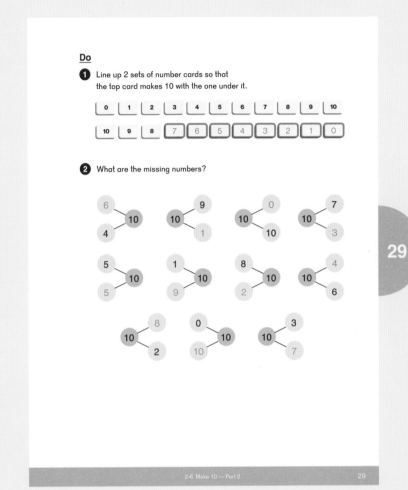

▲ Flash Card Number Bonds

Exercise 6 • page 25

Materials: Number Bonds to 10 Flash Cards (BLM)

Have students practice their facts with Number Bonds to 10 Flash Cards (BLM). Have them see how fast they can give the missing number.

This activity emphasizes speed and is included as a way to practice number bonds to 10 to automaticity.

Teachers may include it as a center/game day activity or provide students with their own sets of cards to practice and quiz themselves at home or during free time.

▲ Pyramid Bonds

Materials: Deck of cards made of Ace (1) to 10 and Queen (0), or use other cards from prior lessons

This is an individual activity. Cards should be laid out in a pyramid format as shown below.

When a player finds two cards that make 10, she can pick them up.

Fill empty spots with cards from the deck.

Continue until all pairs have been found.

　　　Teacher's Guide 1A Chapter 2　　　© 2017 Singapore Math Inc.

Lesson 7 Practice

Objective

- Practice working with number bonds to 10.

After students complete the **Practice** in the textbook, have them continue to compose and split numbers to 10 by playing games and activities from this chapter.

Students should continue to practice all number bonds to 10 until they are transferred to memory. This skill lays the foundation for future work with greater numbers.

Developing fluency with number bonds to 10 may take some students longer than others. Chapter activities can be included as warm-ups or during centers throughout **Chapter 3** and **Chapter 4**.

Activity

▲ Ten Train

Materials: Dominoes

Using a set of dominoes, have students make 10s to connect.

Students turn all dominoes facedown, then each player draws 7 dominoes. One domino from the pile is turned faceup to begin.

Players try to make a 10 with a domino from their hands. If a player can't make a 10, he draws dominoes from the pile until he can play.

The winner is the first person to play all of her dominoes.

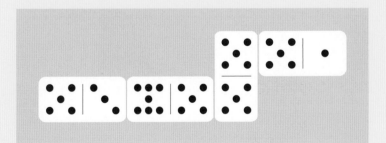

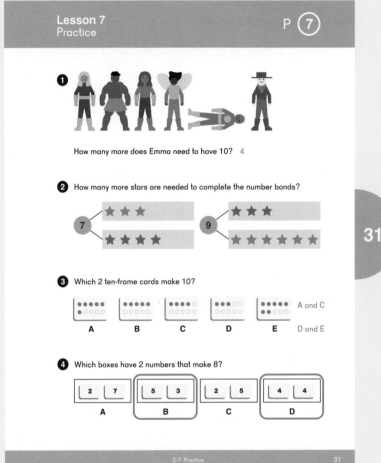

7 Possible student stories:

- 1 balloon has a bow on its string. 9 balloons have no bows.
- 2 balloons are star-shaped. 8 balloons are oval-shaped.
- 3 balloons have spots. 7 balloons have no spots.
- 4 balloons have short strings. 6 balloons have longer strings.
- 5 balloons are pink. 5 balloons are blue.

Brain Works

★ 3-Part Number Bond

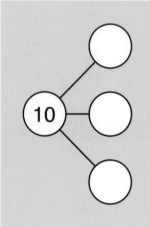

Try including problems like these on the board or in a center.

How many different solutions can you find?

Exercise 7 • page 27

---

5 Find pairs of numbers in each box that make 9.

3 and 6   (a)

| 3 | 2 | 4 |
| 8 | 5 | 6 |

5 and 4

(b)

| 9 | 8 | 2 |
| 4 | 7 | 0 |

9 and 0

7 and 2

6 Complete the number bonds.

8 / 9 / 1   0 / 5 / 5   6 / 8 / 2   2 / 7 / 5

6 / 2 / 4   6 / 9 / 3   9 / 2 / 7   10 / 5 / 5

7 Make up number stories about the balloons. Write a number bond for each.

Answers will vary.

10 / 1 / 9   10 / 2 / 8   10 / 3 / 7   10 / 4 / 6   10 / 5 / 5

Exercise 7 • page 27

32   2-7 Practice

Chapter 2 Number Bonds

Exercise 1

Basics

1 Complete the number bonds.

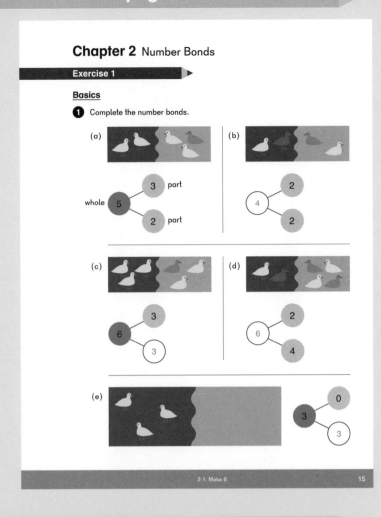

(a)

whole 5 — 3 part / 2 part

(b)

4 — 2 / 2

(c)

6 — 3 / 3

(d)

6 — 2 / 4

(e)

3 — 0 / 3

2-1 Make 6   15

Practice

2 Complete the number bonds.

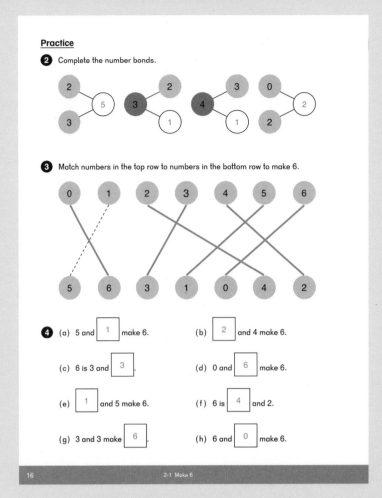

2 — 5 / 3   3 — 2 / 1   4 — 3 / 1   0 — 2 / 2

3 Match numbers in the top row to numbers in the bottom row to make 6.

0 1   2 3   4 5   6

5 6   3 1   0 4   2

4 (a) 5 and [1] make 6.   (b) [2] and 4 make 6.

(c) 6 is 3 and [3].   (d) 0 and [6] make 6.

(e) [1] and 5 make 6.   (f) 6 is [4] and 2.

(g) 3 and 3 make [6].   (h) 6 and [0] make 6.

16   2-1 Make 6

Exercise 2

Basics

1 Draw the missing parts to make 7, then write the missing numbers.

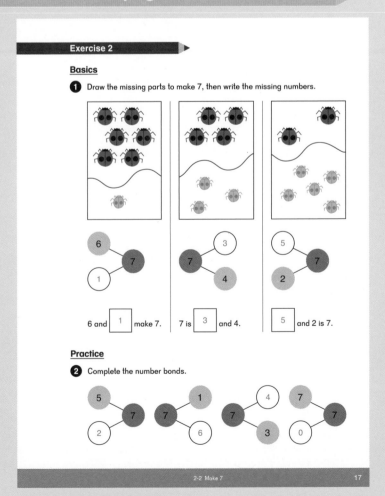

6 — 7 / 1   7 — 3 / 4   5 — 7 / 2

6 and [1] make 7.   7 is [3] and 4.   [5] and 2 is 7.

Practice

2 Complete the number bonds.

5 — 7 / 2   7 — 1 / 6   7 — 4 / 3   7 — 7 / 0

2-2 Make 7   17

3 Color each set of 2 numbers next to each other that make 7.

6	1	3	4	6
5	7	8	3	2
7	4	1	0	9
9	1	4	7	5
2	8	7	5	3
0	6	8	2	6

4 Write numbers to make number bonds.   Answers may vary.

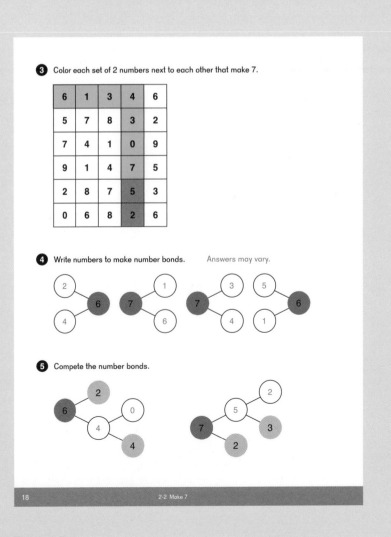

2 — 6 / 4   1 — 7 / 6   7 — 3 / 4   5 — 6 / 1

5 Compete the number bonds.

6 — 2 / 4 — 0 / 4   7 — 5 — 2 / 3 / 2

18   2-2 Make 7

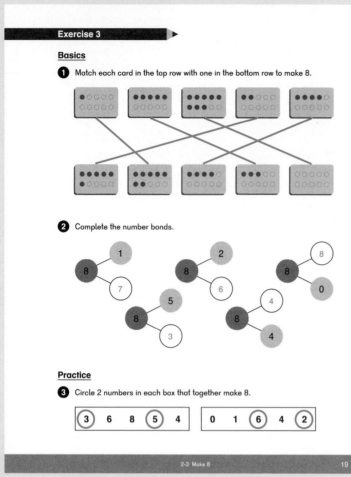

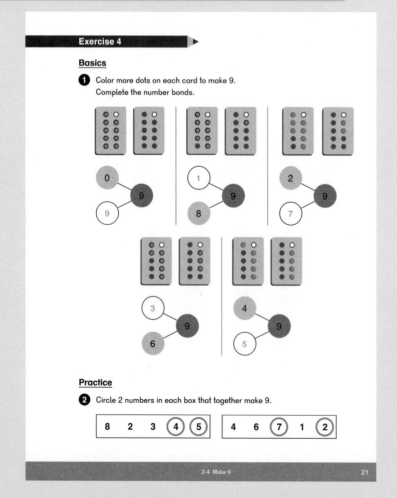

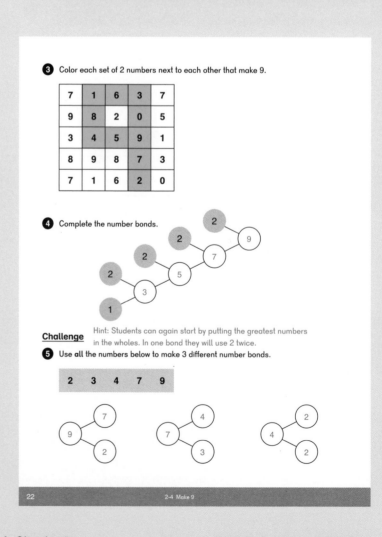

Teacher's Guide 1A Chapter 2

Exercise 5

Basics

❶ Complete the number bonds.

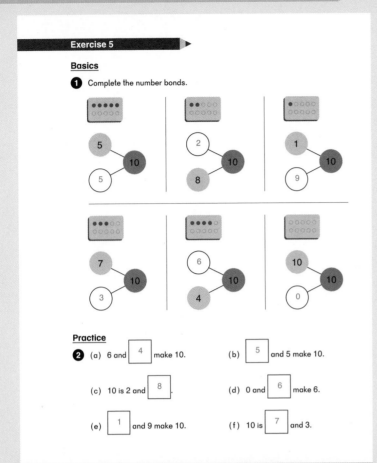

Practice

❷ (a) 6 and [4] make 10.   (b) [5] and 5 make 10.

(c) 10 is 2 and [8].   (d) 0 and [6] make 6.

(e) [1] and 9 make 10.   (f) 10 is [7] and 3.

❸ Circle each set of 2 numbers next to each other that make 10.
Find 10 pairs.

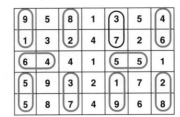

Note: In this puzzle, the there is no overlap of the pairs that make 10, and there are no pairs next to each other diagonally.

❹ Complete the number bonds.

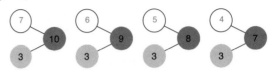

Challenge

Note: From the Challenge in Exercise 4, students should realize they can reuse both 4 and 2.

❺ Use all the numbers below to make 3 different number bonds.

| 2 | 4 | 8 | 10 |

2-5 Make 10 — Part 1   23

24   2-5 Make 10 — Part 1

Exercise 6

Basics

❶ Complete the number bonds.

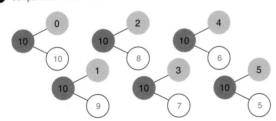

❷ Match a number in the top row with a number in the bottom row to make 10.

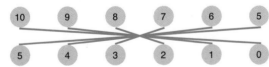

Practice

❸ Circle 2 numbers in each box that together make 10.

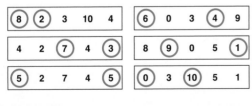

❹ Circle each set of 2 numbers next to each other that make 10.
The corners of the boxes they are in can be touching for them to be next to each other.
Find 10 pairs.

Note: In this puzzle, there is no overlap of the pairs that make 10, but pairs can be next to each other diagonally.

❺ Complete the number bonds.

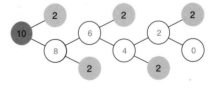

Challenge

❻ Use all the numbers below to make 3 different number bonds.

| 0 | 1 | 10 |

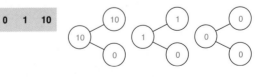

Note: From their experience earlier, students may think to use 0 three times in one bond, though they have not seen this number bond.

2-6 Make 10 — Part 2   25

26   2-6 Make 10 — Part 2

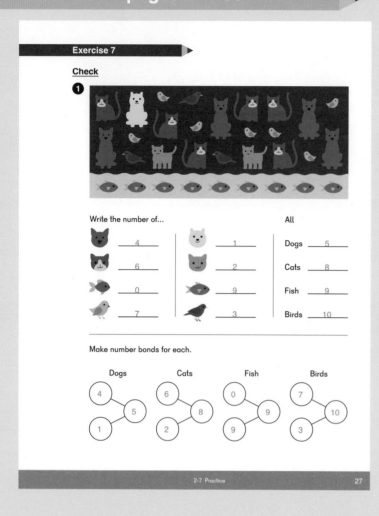

Exercise 7

Check

1

Write the number of...

🐺 _4_   🐱 _1_   All

🐱 _6_   🐱 _2_   Dogs _5_

🐟 _0_   🐟 _9_   Cats _8_

🐦 _7_   🐦 _3_   Fish _9_

Birds _10_

Make number bonds for each.

Dogs   Cats   Fish   Birds

2-7 Practice   27

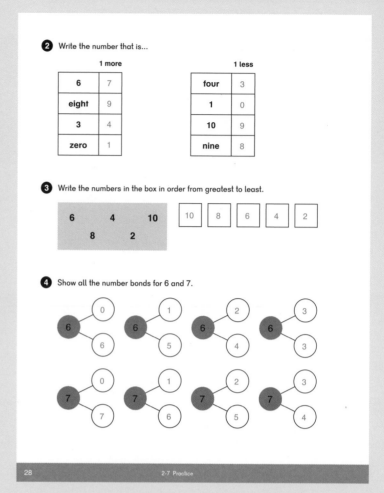

2 Write the number that is...

1 more			1 less	
6	7		four	3
eight	9		1	0
3	4		10	9
zero	1		nine	8

3 Write the numbers in the box in order from greatest to least.

6   4   10   8   2

10   8   6   4   2

4 Show all the number bonds for 6 and 7.

28   2-7 Practice

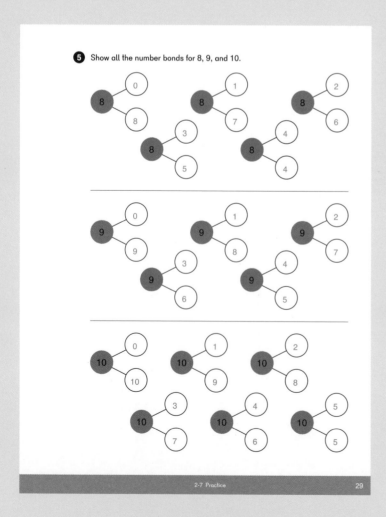

5 Show all the number bonds for 8, 9, and 10.

2-7 Practice   29

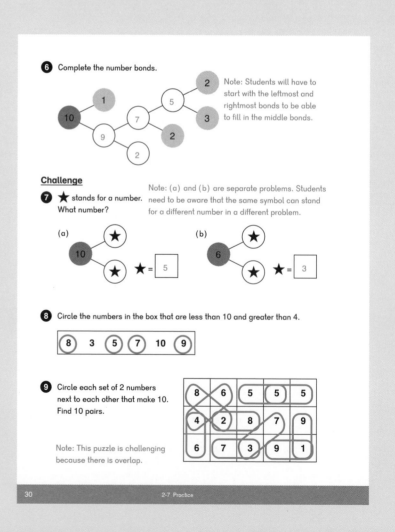

6 Complete the number bonds.

Note: Students will have to start with the leftmost and rightmost bonds to be able to fill in the middle bonds.

Challenge

7 ★ stands for a number. What number?

Note: (a) and (b) are separate problems. Students need to be aware that the same symbol can stand for a different number in a different problem.

(a) 10   ★ = 5

(b) 6   ★ = 3

8 Circle the numbers in the box that are less than 10 and greater than 4.

8   3   ⑤   ⑦   10   ⑨

9 Circle each set of 2 numbers next to each other that make 10. Find 10 pairs.

8	6	5	5	5
4	2	8	7	9
6	7	3	9	1

Note: This puzzle is challenging because there is overlap.

30   2-7 Practice

Suggested number of class periods: 8 – 9

Lesson		Page	Resources		Objectives
	Chapter Opener	p. 51	TB:	p. 33	Tell addition stories.
1	Addition as Putting Together	p. 52	TB: WB:	p. 34 p. 31	Understand addition as "putting together." Solve addition problems.
2	Addition as Adding More	p. 55	TB: WB:	p. 37 p. 33	Understand addition as "adding to." Solve addition problems.
3	Addition with 0	p. 57	TB: WB:	p. 39 p. 35	Add 0 to a number or add a number to 0.
4	Addition with Number Bonds	p. 59	TB: WB:	p. 41 p. 37	Use the part-whole number bond relationship to add two numbers within 10.
5	Addition by Counting On	p. 62	TB: WB:	p. 44 p. 41	Add 1, 2, or 3 to a number within 10 by counting on.
6	Make Addition Stories	p. 64	TB: WB:	p. 46 p. 43	Tell addition stories given an addition situation.
7	Addition Facts	p. 67	TB: WB:	p. 49 p. 45	Learn the addition facts to 10 to mastery.
8	Practice	p. 69	TB: WB:	p. 51 p. 49	Practice addition facts.
	Workbook Solutions	p. 73			

This chapter builds on **Chapter 2: Number Bonds** by helping students progress from the number bond format to formal addition with symbols.

Lessons will generally include concrete, pictorial, and abstract representations. It is important that teachers include the concrete portion of a lesson as many students have memorized some basic skills but have not developed a true understanding of concepts.

For the addition sign, the two lines should be straight and of equal length. By going across horizontally first, students are more likely to write the addition sign neatly.

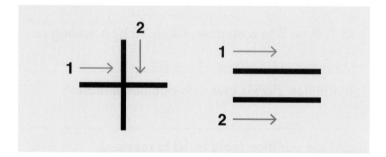

In this chapter, students will use the term "equation" for an addition sentence. The language of number bonds, "3 and 5 make 8." is changed to the "3 **plus** 5 **equals** 8."

Students will be shown a variety of number bond forms and orientations. This variety helps students to realize the sum is not in a given location, but is the whole.

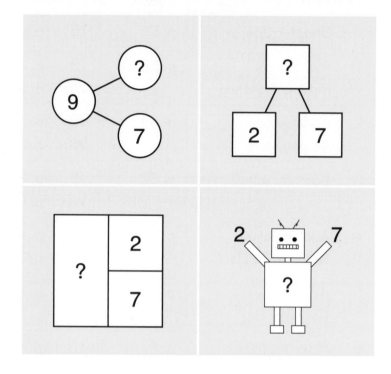

Students also need a large variety of examples to recognize that the important quality of a number bond is the part-whole relationship, not that numbers are written in circles, or large numbers go on top, etc.

Note on terminology: An "addition equation" is the formal equation using symbols, read as "3 plus 5 equals 8," as opposed to, "3 and 5 make 8."

Teacher's Guide 1A Chapter 3

Materials

- Linking cubes
- Sticky notes (2 different colors)
- Dry erase markers
- Two-color counters
- Dry erase sleeves
- Counters in a variety of shapes and colors
- Cups or buckets
- Blank dice
- Regular dice
- Game board
- Sidewalk chalk
- Paper plates
- Painter's tape
- Picture book or magazine
- Index cards or construction paper
- 10-sided die
- Whiteboards

Note: Materials for Activities will be listed in detail in each lesson.

Blackline Masters

- Number Bond Story Template
- Flip and Count Addition
- Equation Symbol Cards
- Adding Zero Alligator Cards
- Roll and Add
- Addition to 10 Cover-up Cards
- Number Cards
- Addition Story Template
- Ten-frame Cards
- Spinner

Storybooks

- *Quack and Count* by Keith Baker
- *Ten Friends* by Bruce Goldstone
- *Domino Addition* by Lynette Long
- *Ten for Me* by Barbara Mariconda
- *Ten Monkey Jamboree* by Dianne Ochiltree
- *Mice Mischief: Math Facts in Action* by Caroline Stills
- *One More Bunny: Adding from One to Ten* by Rick Walton

Letters Home

- Chapter 3 Letter

Notes

Chapter Opener

Objective

- Tell addition stories.

Ask students, "How many items are on each shelf? Can you make a number bond for each set of objects on a shelf?"

Students should observe that:

- 3 mugs are blue and 2 are red.
- There are 4 small plates and 3 large plates.
- 3 bowls have dots on them and 3 bowls are solid-colored.
- There are 5 tall glasses and 3 short glasses.

This lesson may continue with the activity or continue to **Lesson 1**: **Addition as Putting Together**.

Activity

▲ Count and Add the Classroom

Have students look around the classroom to see if they can create a number story using classroom objects. Have them share their number bonds and number stories.

Model a story for students including two parts and the whole. Here are some examples:

- There are 4 tall bookshelves and 1 small bookshelf. There are 5 bookshelves in all.
- There are 4 red chairs and 3 blue chairs at the front table. There are 7 chairs in all.
- ★ · There are 2 l's in Gabriella's first name and 1 in her last name. Her name has 3 l's altogether.

Chapter 3

Addition

There are 3 blue mugs.
There are 2 red mugs.
There are 5 mugs altogether.

What other number stories can you make?

33

33

Lesson 1 Addition as Putting Together

Objectives

- Understand addition as "putting together."
- Solve addition problems.

Lesson Materials

- Linking cubes or attribute bears, 10 each of 2 different colors per student
- Sticky notes, 2 colors
- For struggling students:
  - Number Cards (BLM) 0 to 10
  - Number Bond Story Template (BLM)

Think

Have students model and solve the number story for the birds with linking cubes and write the number bond for the number story.

Learn

Have the students show 5 cubes and 3 cubes, then put them together to see that there are 8 cubes or birds in all. In the past, students have used the terms, "5 and 3 make 8."

Tell students that when we are joining numbers together, we are **adding**. We can use a number bond to help us make an **addition** sentence or **equation**.

Using two of the same color sticky notes, write the parts from the birds number story and put them in the parts on a number bond on the board.

Write the total on the other color of sticky note and put it in the whole.

Introduce the addition symbol and move the sticky notes to show how the parts relate to the story and the parts of an addition equation.

Introduce the equal sign. Move the sticky note from the whole in the number bond to after the equal sign. Note that students have used the + and = sign in

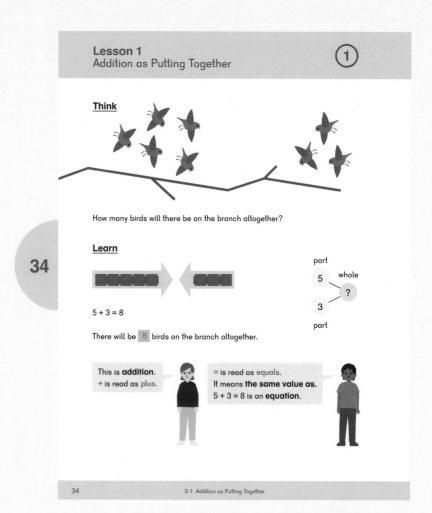

Dimensions Math® Kindergarten B materials and may be familiar with the signs. The term "equation" is new here.

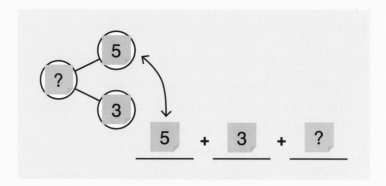

Discuss the bird problem in the textbook and Emma and Alex's comments about equations.

Tell students that the value of 5 + 3 is 8, so we use an equal sign to show that their values are the same.

Tell students that the equal sign (or symbol) means the value of what is on both sides of it is the same.

Write, "8 = 5 + 3" and ask students if this is a **true** equation. It is, because the value on both sides of the = symbol are the same.

Provide students with some other examples of number bonds and have them practice writing addition equations on their whiteboards.

Struggling students can use either sticky notes in two colors or Number Cards (BLM) 0 to 10. They can move the numbers from a Number Bond Story Template (BLM) to an addition equation on their whiteboards.

Do

Students can use linking cubes to model the problems if needed.

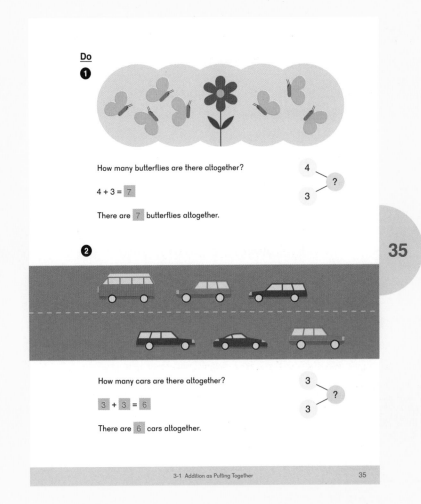

Do

❶

How many butterflies are there altogether?

4 + 3 = 7

There are 7 butterflies altogether.

4
?
3

❷

How many cars are there altogether?

3 + 3 = 6

There are 6 cars altogether.

3
?
3

3-1 Addition as Putting Together   35

35

4 Note that all number bonds are represented by the picture on page 36.

- 5 + 3 = 8; 5 boys and 3 girls make 8 soccer players altogether.
- 2 + 6 = 8; 2 players with glasses and 6 without glasses make 8 soccer players altogether.
- 1 + 7 = 8; 1 player with a headband and 7 without headbands make 8 soccer players altogether.
- 4 + 4 = 8; 4 players in green jerseys and 4 in blue jerseys make 8 soccer players altogether.

Students may also note 8 + 0 = 8:

- 8 soccer players and 0 not playing soccer make 8 soccer players altogether.

Help students recall that they can only add like units — players to players. If students use other terms, they must find a common unit: 0 parents plus 8 soccer players is 8 people in all.

Activity

▲ Flip and Count Addition

Materials: Flip and Count Addition (BLM), 10 two-color counters per student, dry erase sleeves, dry erase markers

Students shake up to 10 counters and put them on the Flip and Count Addition (BLM) worksheet in the correct parts.

Students fill in _____ + _____ = _____.

3 There are 5 red crayons. There are 5 yellow crayons. How many crayons are there altogether?

5 + 5 = 10

There are 10 crayons altogether.

4

Write addition equations for this picture.
Answers will vary.

Exercise 1 • page 31

36   3-1 Addition as Putting Together

36

Exercise 1 • page 31

Teacher's Guide 1A Chapter 3

Lesson 2 Addition as Adding More

Objectives

- Understand addition as "adding to."
- Solve addition problems.

Lesson Materials

- Linking cubes, 10 of one color per student
- Counters

Think

Have 4 students go to the front of the class and ask the rest of the class how many students there are. (4 students)

Send 3 more students up to the front and ask how many students you sent up. (3 students)

Ask, "How many students are there now?" (7 students)

Ask students to explain what happened. (There were 4 kids up front and 3 more joined. Now there are 7 altogether.)

Point out that something changed. There were some students and some more joined them.

Show students textbook page 37. Read the story and ask students what they notice about the birds. (Part of the birds are on a branch and part of the birds are flying to join them. How many birds will there be in all?)

Have students model the number story for the birds by putting together linking cubes.

Learn

Discuss the number bond and the equation in the textbook.

The activity **Addition Equations** on the following page provides practice putting the + and = signs in the equation along with practicing number combinations to 10.

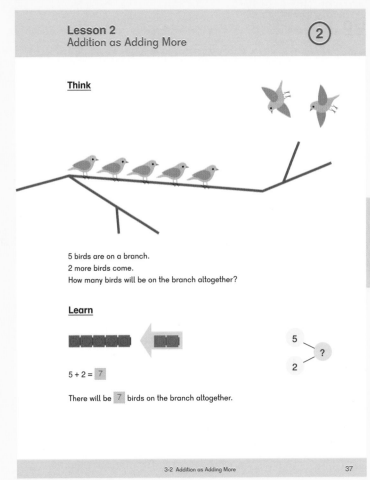

Do

3 Students may write any of the equations:

$$6 + 2 = 8$$
$$2 + 6 = 8$$
$$8 = 2 + 6$$
$$8 = 6 + 2$$

Activity

▲ Addition Equations

Materials: Equation Symbol Cards (BLM)
(Just + and =), 10 counters per pair of students

Player 1 creates an addition equation with counters
and symbols on a whiteboard.

Player 2 tells the answer.

Players can count to check.

Exercise 2 • page 33

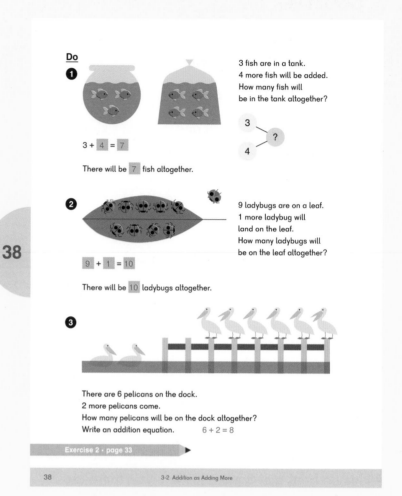

Do

1 3 fish are in a tank.
4 more fish will be added.
How many fish will
be in the tank altogether?

$3 + \boxed{4} = 7$

There will be $\boxed{7}$ fish altogether.

2 9 ladybugs are on a leaf.
1 more ladybug will
land on the leaf.
How many ladybugs will
be on the leaf altogether?

$\boxed{9} + \boxed{1} = \boxed{10}$

There will be $\boxed{10}$ ladybugs altogether.

3 There are 6 pelicans on the dock.
2 more pelicans come.
How many pelicans will be on the dock altogether?
Write an addition equation. $6 + 2 = 8$

Exercise 2 • page 33

38   3-2 Addition as Adding More

Lesson 3 Addition with 0

Objective

- Add 0 to a number or add a number to 0.

Lesson Materials

- 2 cups or buckets
- 7 small items from classroom
- 2 plates

Think

Put 7 objects (blocks, etc.) into one cup and leave the other empty. Ask students what happens when we try to add zero to another number. Does the other number change? And what happens when we add a number to zero?

Have a plate filled with counters and an empty plate to illustrate both scenarios.

Discuss student ideas and equations.

Learn

Show students the number bond for 7 and 0. Have them write the addition equation on their whiteboards.

Read and discuss the **Think** and **Learn** problem with the birds on the branch.

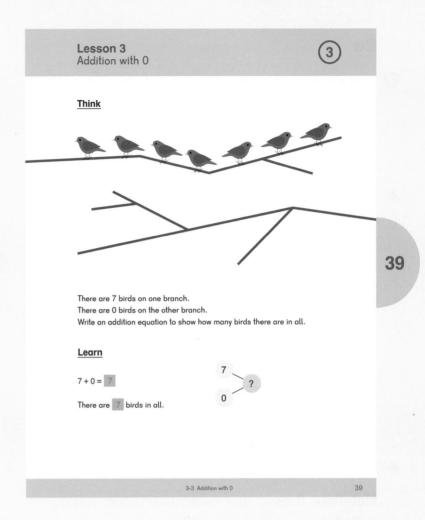

Do

3 This question provides examples of equations where 0 is in different locations.

Activities

▲ Alligator! Alligator! Alligator!

Materials: Adding Zero Alligator Cards (BLM)

Show Adding Zero Alligator Cards (BLM) to class as a whole group. If you give them the cue, "Think, 2, 1, answer," they should answer chorally. When an alligator comes up, students shout, "Alligator! Alligator! Alligator!"

While the game seems goofy and may be loud, it keeps all students engaged because they do not want to miss the alligator card when it comes up.

▲ Roll and Add

Materials: 2 blank dice modified with faces: 0, 1, 2, 3, 4, 5 (use blank dice, put stickers over regular dice, or use wooden cubes to create customized dice), Roll and Add (BLM)

Students roll two dice. They write an addition equation created by using the number shown on each roll as a part on a whiteboard.

Students may record their answers on the Roll and Add (BLM).

Exercise 3 • page 35

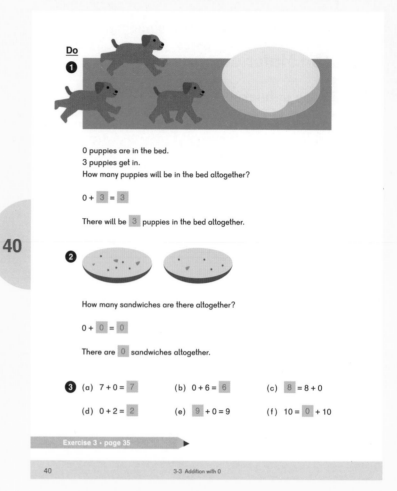

Do

1

0 puppies are in the bed.
3 puppies get in.
How many puppies will be in the bed altogether?

$0 + \boxed{3} = \boxed{3}$

There will be $\boxed{3}$ puppies in the bed altogether.

2

How many sandwiches are there altogether?

$0 + \boxed{0} = \boxed{0}$

There are $\boxed{0}$ sandwiches altogether.

3 (a) $7 + 0 = \boxed{7}$ (b) $0 + 6 = \boxed{6}$ (c) $\boxed{8} = 8 + 0$

    (d) $0 + 2 = \boxed{2}$ (e) $\boxed{9} + 0 = 9$ (f) $10 = \boxed{0} + 10$

Exercise 3 • page 35

40   3-3 Addition with 0

Objective

- Use the part-whole number bond relationship to add two numbers within 10.

Lesson Materials

- Linking cubes, 10 each of 2 different colors per student
- Two-color counters

Think

Ask students, "Can we write an addition equation in more than one way?"

Show students the **Think** task and have them represent the birds with two different color linking cubes. Discuss what equations they can find using the cubes to represent the birds.

Guide them to see that 3 is one part and 4 is another part. Together, both parts make the whole and finding the whole is the same as finding the total or sum.

4 + 3 and 3 + 4 both have the same total of 7.

Learn

Write the two number bonds on the board or look at the textbook pictures.

Ask students:

- What is Sofia thinking?
- Is it true that 4 + 3 equals 3 + 4?
- Is it also true that 5 + 4 equals 4 + 5?

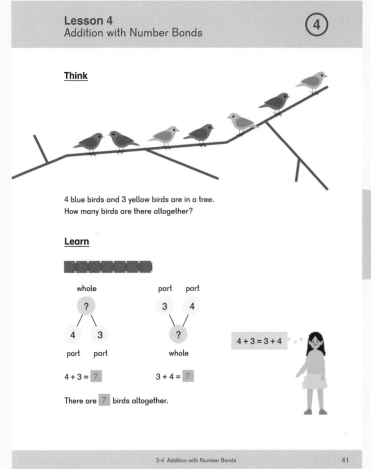

Lesson 4
Addition with Number Bonds (4)

41

Think

4 blue birds and 3 yellow birds are in a tree. How many birds are there altogether?

Learn

whole   part   part
?   3   4
4   3   ?
part part whole

4 + 3 = 7   3 + 4 = 7

4 + 3 = 3 + 4

There are 7 birds altogether.

3-4 Addition with Number Bonds   41

Do

5 This question on page 43 introduces a new concept: finding a missing addend in an equation.

Encourage students to relate a missing part to a missing addend.

Activities

● **Make Addition Equations**

Materials: Several sets of Number Cards (BLM) 0 to 10, Equation Symbol Cards (BLM)

This individual activity requires the following Number Cards (BLM) and Equation Symbol Cards (BLM):

2, 2, 3, 4, 5, 6, +, +, =, =

▲ Using all these cards, can you make two addition equations?

Students may need to try different equations to solve the problem correctly:

2 + 3 = 5   4 + 2 = 6

★ For more challenge, add the cards 3, 4, 7, +, = (or 5, 1, 6, +, =, or create your own) and have students make three equations.

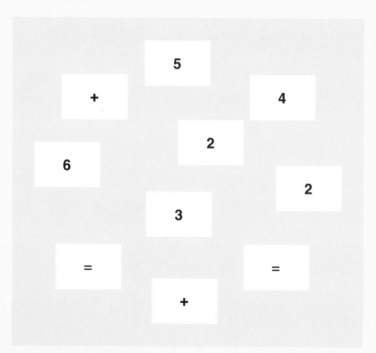

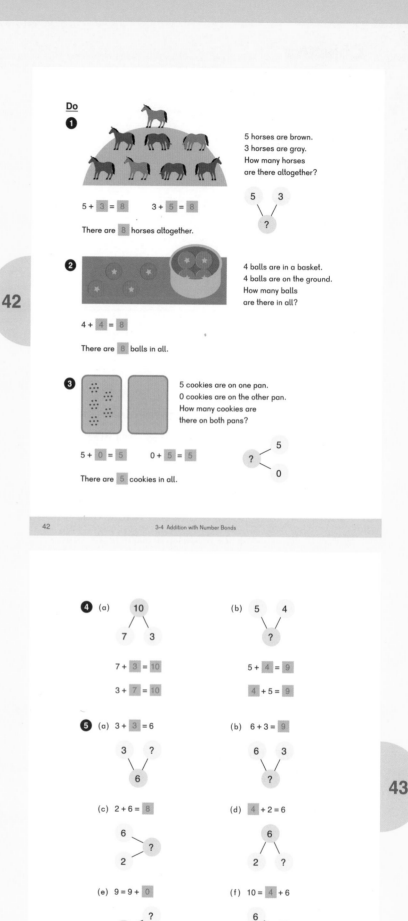

Exercise 4 · page 37

▲ Cover-up

Materials: Addition to 10 Cover-up Cards (BLM)

Shuffle Addition to 10 Cover-up Cards (BLM) and show them to students, using your finger or a strip of paper to cover one of the addends or the total. Have students tell you the number that is covered up.

▲ Make a Match

Materials: Addition to 10 Cover-up Cards (BLM)

This is a small group activity that plays like **Concentration** and the **Make a Match** game in **Chapter 1: Lesson 2** on page 10 of this Teacher's Guide.

Have students lay the Addition to 10 Cover-up Cards (BLM) facedown and take turns turning over two cards. If the two cards sum to the same whole (or total), the student keeps those cards.

For example, if Player 1 draws:

$3 + 4 = 7$ $2 + 5 = 7$

She has a match and keeps the cards.

If Player 2 draws:

$3 + 2 = 5$ $2 + 4 = 6$

He does not have a match, so he will return the cards to the pile and play will continue.

Exercise 4 • page 37

Lesson 5 Addition by Counting On

Objective

- Add 1, 2, or 3 to a number within 10 by counting on.

Lesson Materials

- Counters (bears, cubes, etc.), 10 for teacher and each student
- 2 paper plates for teacher and each student

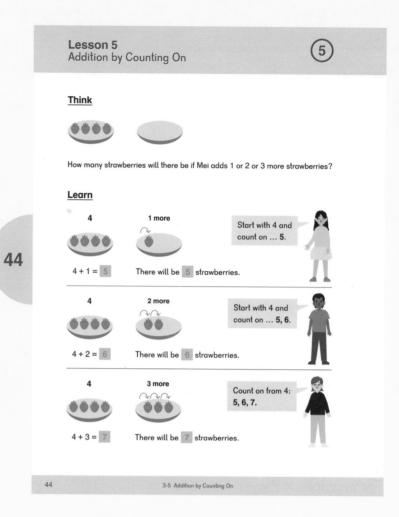

Think

Pose the **Think** problem with the strawberries and provide students with counters and paper plates. Have students work the problem as it is being demonstrated. Explain to students we can add by counting.

Add 1 counter to the second plate and say, "5." Add another and say, "6," then add another and say, "7." Ask, "How many counters did we add?" On the board, write, "4 + 3 = 7."

Repeat, starting with different numbers and have students count aloud as you put more counters on the plate. Stop after placing up to 3 more counters on the plate and have students tell how many were added and how many there are altogether.

Write some additional equations and have students solve by counting on. Provide one or two examples where the greater number comes second, for example, 2 + 5. Point out to students that when we are counting on, it is easier to start with the greater number: 5, 6, 7.

Learn

Discuss textbook page 44. Ask if they can form an equation if no strawberries were added to the second plate.

Do

2 Have students start with 6 and count on 3.

3 Allow students who are struggling to use counters or other manipulatives.

Activities

▲ Counting-on Dice

Materials: 1 regular die, 1 die with the faces labeled 1, 1, 2, 2, 3, 3, game board and markers

This can be used as a small group or partner activity. Almost any game board where students need to count on to move their pieces will work for this activity. Have students roll both the regular and modified dice together, and determine how many spaces to move by adding them together using the strategy of counting on.

Players take turns rolling the two dice and adding the numbers together.

▲ Counting-on Game Board

Materials: 1 regular die, 1 die with the faces labeled 1, 1, 2, 2, 3, 3, posterboard, markers

Have students create their own game board with spaces for 0 to 10. Playing in pairs, students take turns rolling a die with faces showing only 1, 2, and 3 (from **Counting-on Dice**) and moving a marker along their own game board. The first player past 10 wins.

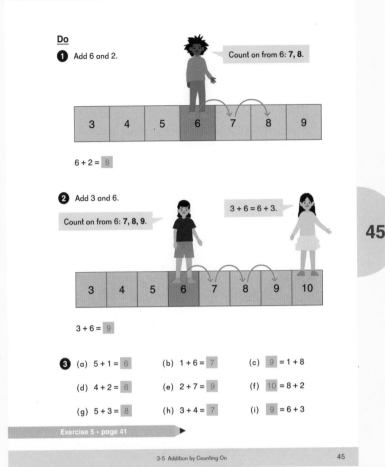

▲ Counting-on Hop

Materials: Chalk or painter's tape, 1 regular die, 1 die with the faces labeled 1, 1, 2, 2, 3, 3

This activity helps students see that you don't count the number you start on. Create a large number line outside with chalk or inside with painter's tape. Using the modified die from **Counting-on Dice**, students take turns being the Dice Roller and the Hopper. The Hopper starts on 0 and hops the number that the Dice Roller rolls. When they get to 10 or beyond, players switch roles.

Exercise 5 • page 41

Lesson 6 Make Addition Stories

Objective

- Tell addition stories given an addition situation.

Lesson Materials

- Picture book from classroom, or nearly any children's book or magazine

Think

Have students discuss the **Think** task. Ask students to make up addition stories for the rabbits.

Learn

Draw the number bond on the board and discuss Dion's story in relation to the number bond. Ask students what other addition stories they can make that will be able to use the same number bond. Possible student responses for a total of 6 include:

- 2 rabbits are hopping and 4 rabbits are not hopping.
- 2 rabbits have white tails. 4 do not.
- 2 rabbits have floppy ears. 4 rabbits have stand-up ears.
- 3 rabbits are brown and 3 are gray.
- 1 rabbit has a carrot. 5 rabbits do not have a carrot.

Write the addition equations as students tell the stories.

While the sentences in the stories all have a specific structure to encourage understanding, students need not imitate it.

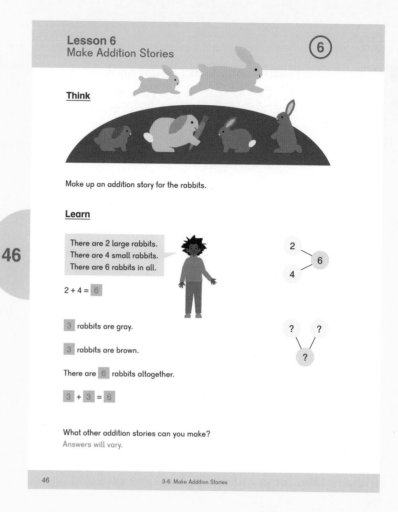

46

Lesson 6
Make Addition Stories   6

Think

Make up an addition story for the rabbits.

Learn

There are 2 large rabbits.
There are 4 small rabbits.
There are 6 rabbits in all.

$2 + 4 = 6$

2
4
6

3 rabbits are gray.

3 rabbits are brown.

There are 6 rabbits altogether.

$3 + 3 = 6$

? ?
?

What other addition stories can you make?
Answers will vary.

46   3-6 Make Addition Stories

Do

1 Have students share their stories. Possible student answers:

(a) Oranges

- There are 5 oranges in a bowl. 3 oranges are in a bag. There are 8 oranges altogether.
- 5 oranges are darker colored. 3 oranges are lighter colored.
- 3 oranges have leaves. 5 oranges do not have leaves.

(b) Books

- There are 3 green books and 3 pink books. There are 6 books in all.
- 3 books are on the shelf and 3 books are not on the shelf.
- 3 are big books and 3 are smaller books.
- 3 books have purple on the edge (spine) and 3 books do not have any purple.

(c) Pencils

- There are 2 yellow pencils and 7 pencils that aren't yellow. There are 9 pencils in all.
- 7 are long pencils. 2 are short pencils.
- 7 pencils do not have erasers. 2 pencils have erasers.

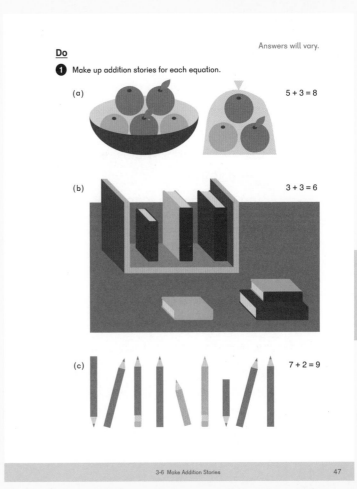

47

2 Have students use whiteboards to write and show their equations.

(a) Children (2 + 3 = 5 is given as an example)
- There is 1 child with glasses and 4 without glasses. There are 5 children in all.
- 2 children are wearing yellow shirts and 3 are wearing red shirts.
- 4 children are wearing long pants. 1 child is wearing shorts.
- 3 children are wearing caps. 2 children are not.

(b) Flowers
- There are 5 tall flowers and 2 short flowers. There are 7 flowers altogether.
- 1 flower has an orange middle and 6 have yellow middles.
- There are 4 purple flowers and 3 red flowers.
- There are 7 flowers in one planter and 0 flowers in the other planter.

(c) Turtles
- There are 4 green turtles and 4 brown turtles. There are 8 turtles in all.
- 3 turtles are in the water. 5 turtles are on land.
- 1 turtle is eating a leaf. 7 turtles are not eating.
- 2 turtles have pink spots. 6 turtles have blue spots.

Activities

▲ Number Stories from Pictures

Almost any storybook or children's magazine lends itself to making number stories from pictures.

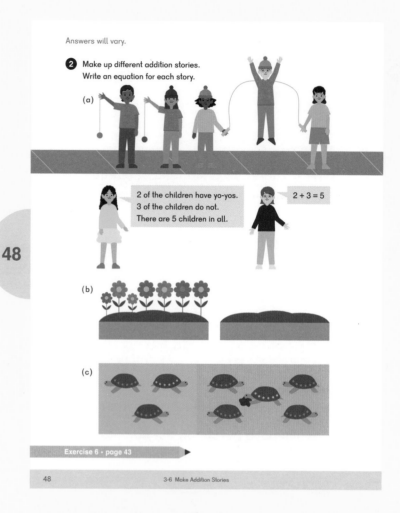

Answers will vary.

2 Make up different addition stories. Write an equation for each story.

(a)

2 of the children have yo-yos. 3 of the children do not. There are 5 children in all.

2 + 3 = 5

(b)

(c)

Exercise 6 • page 43

48   3-6 Make Addition Stories

▲ Draw an Addition Story

Materials: Addition Story Template (BLM)

As in the **Chapter Opener** in **Chapter 2**, have students create and illustrate number stories using the Addition Story Template (BLM). They should show the number bonds and addition equations of their stories. They do not need to write their stories out in words.

Students can share their number stories.

◀ **Exercise 6 • page 43**

Lesson 7 Addition Facts

Objective

- Learn the addition facts to 10 to mastery.

Lesson Materials

- Index cards, 50 per student, or construction paper
- Number Cards (BLM) 1 to 10, 1 set per student
- Ten-frame Cards (BLM) 1 to 10
- 10-sided die or Spinner (BLM)
- Paper clip for Spinner

Many students will already have mastered the addition facts to 10. The teacher should assess students on these facts.

Think

Provide students with index cards and have them create their own flash cards like those shown on page 49 for future practice and games. They can lay them out in a similar pattern to the textbook illustration while looking for patterns.

Students can also fold construction paper into eight equal parts and cut out their own flash cards.

Have students find the answers and look for patterns in the fact cards when organized as on page 49. There are many different patterns students may notice.

To extend, have students find different ways of organizing their cards based on different patterns. For example, they could line up all the sums to 10 on the left, then the 9s, and so on.

Learn

Have students lay out Number Cards (BLM) 1 to 10 and sort their flash cards under each number.

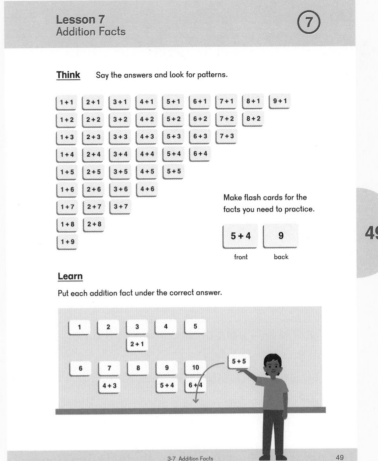

© 2017 Singapore Math Inc.

Do

1 Mei is thinking about the problem as 5 dots on the top row and 4 dots on the bottom row, or 5 + 4 = 9. Students may also see:

- 1 less than 10 is 9. 10 − 1 = 9
- 4 pairs of dots on the left and 1 dot alone on the right. 8 + 1 = 9
- 4 dots in the bottom row is 1 less than 5 dots in the top row. 5 − 1 = 4

Alex is thinking about the problem as 5 dots on the top row and 2 dots on the bottom row, or 5 + 2 = 7. Students may also see:

- A group of 4 dots and a group of 3 dots. 4 + 3 = 7
- 3 less than 10 dots. 10 − 3 = 7
- 5 dots in the top row is 3 more than 2 dots in the bottom row. 5 − 3 = 2

To extend, teachers can show students Ten-frame Cards (BLM) and ask, "What is 2 more than this number? What is 4 more than this number?"

2 Students play in pairs with a 10-sided die or Spinner (BLM).

Exercise 7 • page 45

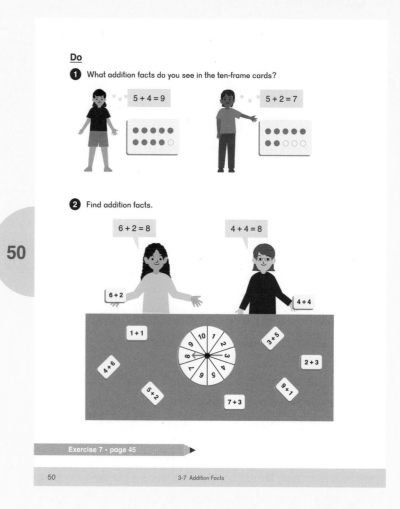

50

Objective

- Practice addition facts.

Practice

After students complete the **Practice** in the textbook, have them continue adding numbers to 10 by playing games from this chapter.

Do

2 Possible stories include:

- There is 1 flower with a pink middle. There are 9 flowers with brown middles. There are 10 flowers in all. $1 + 9 = 10$
- There are 2 with leaves and 8 without leaves. $2 + 8 = 10$
- There are 3 small flowers and 7 big flowers. There are 10 flowers in all. $3 + 7 = 10$
- 4 flowers are in the orange can. 6 flowers are in the blue can. There are 10 flowers in all. $4 + 6 = 10$
- 5 are red flowers. 5 are yellow flowers. There are 10 flowers in all. $5 + 5 = 10$

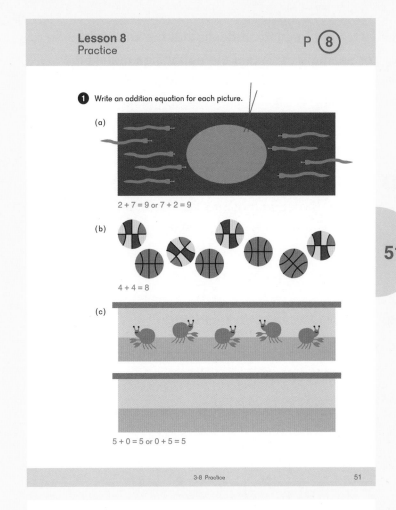

Lesson 8
Practice   P 8

1 Write an addition equation for each picture.

(a)
$2 + 7 = 9$ or $7 + 2 = 9$

(b)
$4 + 4 = 8$

(c)
$5 + 0 = 5$ or $0 + 5 = 5$

3-8 Practice   51

51

2 Make up addition stories about the flowers. Write an equation for each.

Answers will vary.

52

52   3-8 Practice

6 (a) – (f) Remind students that the value of 4 + 5, for example, is 9, and the equal sign means the value of what is on both sides of it is the same. (Likewise, 9 = 9 is also a true equation.) Prompt with questions like, "What number added to 7 will give us the same value as 4 + 5?"

7 There should be discussion similar to **6** identifying whether the equations are true or false. Have students explain their reasoning.

Activity

▲ **Which One Doesn't Belong?**

There could be multiple reasons why each number may not belong.

Ask students which number doesn't belong.

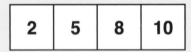

A few possible answers:

$\boxed{2}$ doesn't belong because it's the least. It has a question mark as part of it's shape.

$\boxed{5}$ doesn't belong because it's an odd number.

$\boxed{8}$ doesn't belong because it's the only number without a straight line as part of it.

$\boxed{10}$ doesn't belong because it is the only two-digit number.

$\boxed{5}$ doesn't belong because it is not part of an "8 and 2 make 10" number bond.

★ $\boxed{10}$ does not belong because 5 is 3 more than 2, and 8 is 3 more than 5, but 10 is not 3 more than 8.

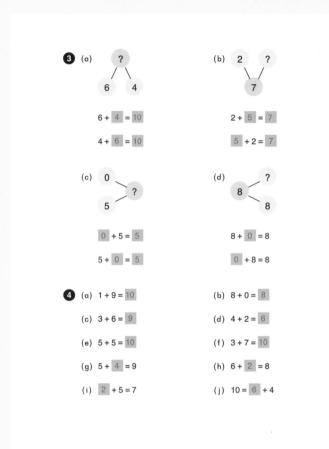

3 (a) ? / 6 4
6 + $\boxed{4}$ = 10
4 + $\boxed{6}$ = 10

(b) 2 ? / 7
2 + $\boxed{5}$ = 7
$\boxed{5}$ + 2 = 7

(c) 0 \ ? / 5
$\boxed{0}$ + 5 = 5
5 + $\boxed{0}$ = 5

(d) ? / 8 \ 8
8 + $\boxed{0}$ = 8
$\boxed{0}$ + 8 = 8

4 (a) 1 + 9 = $\boxed{10}$
(b) 8 + 0 = $\boxed{8}$
(c) 3 + 6 = $\boxed{9}$
(d) 4 + 2 = $\boxed{6}$
(e) 5 + 5 = $\boxed{10}$
(f) 3 + 7 = $\boxed{10}$
(g) 5 + $\boxed{4}$ = 9
(h) 6 + $\boxed{2}$ = 8
(i) $\boxed{2}$ + 5 = 7
(j) 10 = $\boxed{6}$ + 4

5

There are 5 broccoli in the bowl.
There are 3 broccoli outside the bowl.
How many broccoli are there altogether?

$\boxed{5}$ + $\boxed{3}$ = $\boxed{8}$

There are $\boxed{8}$ broccoli altogether.

6 (a) 2 + 8 = 8 + $\boxed{2}$
(b) 4 + 5 = 7 + $\boxed{2}$
(c) 6 + 1 = $\boxed{0}$ + 7
(d) 5 + $\boxed{3}$ = 7 + 1
(e) 3 + $\boxed{4}$ = 1 + 6
(f) $\boxed{3}$ + 2 = 0 + 5

7 Which equations are true?

(a) 3 + 4 = 10
(b) ⬭ 9 = 3 + 6
(c) 3 + 3 = 4 + 4
(d) ⬭ 3 + 6 = 6 + 3
(e) 4 + 1 = 5 + 2
(f) ⬭ 3 + 5 = 2 + 6
(g) ⬭ 3 + 4 = 5 + 2 = 6 + 1
(h) 1 + 2 = 3 + 4 = 7

Exercise 8 • page 49 ▶

54   3-8 Practice

Brain Works

★ Shape Math

Try including problems like these on the board or in a center.

$$\triangle = 0 \qquad \blacksquare = 2 \qquad \bullet = 5$$

Fill in the blanks with the correct totals.

$\triangle + \blacksquare = \underline{\quad}$ $\bullet + \triangle = \underline{\quad}$

$\blacksquare + \bullet = \underline{\quad}$ $\triangle + \triangle = \underline{\quad}$

$\blacksquare + \blacksquare = \underline{\quad}$ $\bullet + \bullet = \underline{\quad}$

Exercise 8 • page 49

Notes

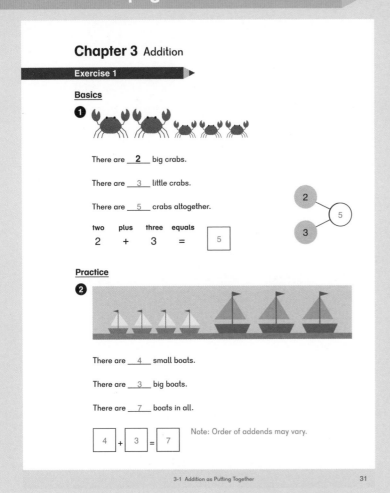

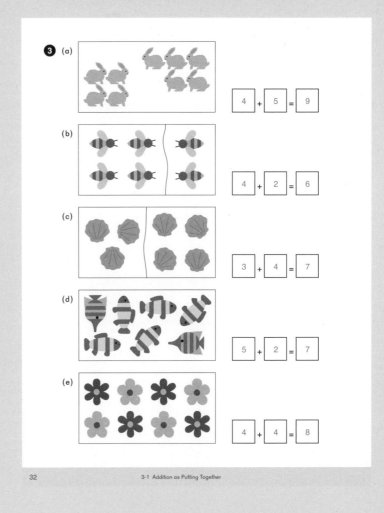

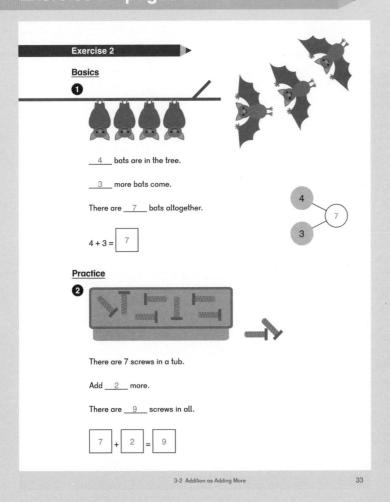

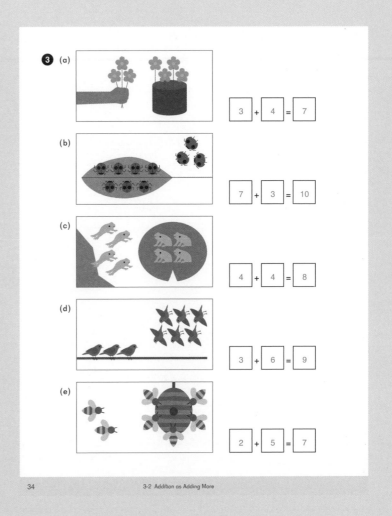

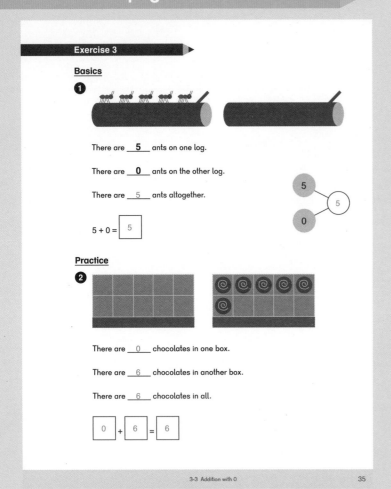

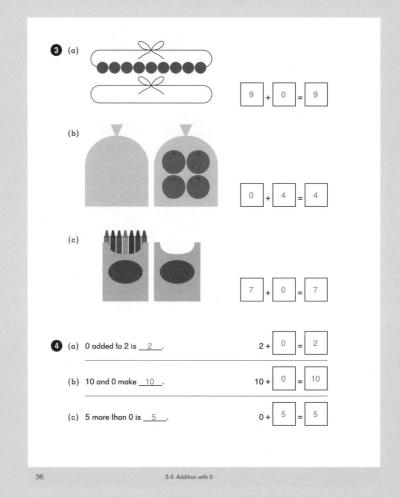

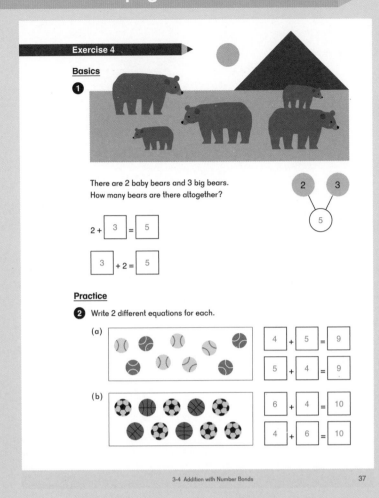

Exercise 4

Basics

① There are 2 baby bears and 3 big bears.
How many bears are there altogether?

$2 + \boxed{3} = \boxed{5}$

$\boxed{3} + 2 = \boxed{5}$

Practice

② Write 2 different equations for each.

(a)
$\boxed{4} + \boxed{5} = \boxed{9}$
$\boxed{5} + \boxed{4} = \boxed{9}$

(b)
$\boxed{6} + \boxed{4} = \boxed{10}$
$\boxed{4} + \boxed{6} = \boxed{10}$

3-4 Addition with Number Bonds   37

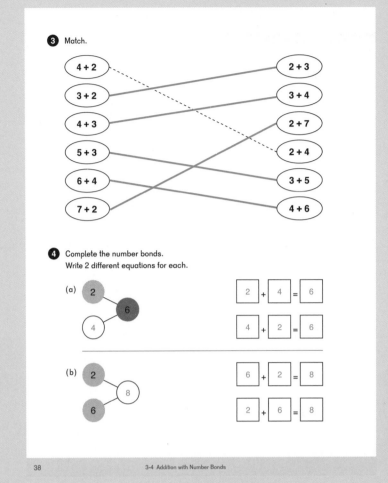

③ Match.

4 + 2
3 + 2
4 + 3
5 + 3
6 + 4
7 + 2

2 + 3
3 + 4
2 + 7
2 + 4
3 + 5
4 + 6

④ Complete the number bonds.
Write 2 different equations for each.

(a) 2, 6, 4
$\boxed{2} + \boxed{4} = \boxed{6}$
$\boxed{4} + \boxed{2} = \boxed{6}$

(b) 2, 8, 6
$\boxed{6} + \boxed{2} = \boxed{8}$
$\boxed{2} + \boxed{6} = \boxed{8}$

38   3-4 Addition with Number Bonds

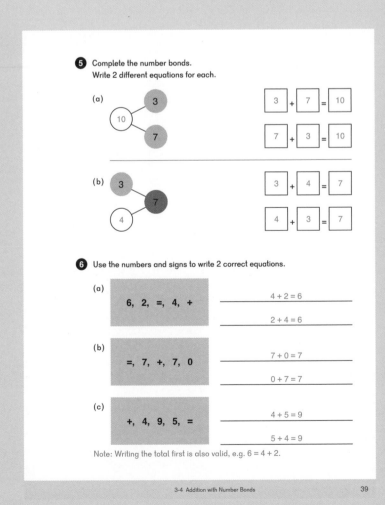

⑤ Complete the number bonds.
Write 2 different equations for each.

(a) 10, 3, 7
$\boxed{3} + \boxed{7} = \boxed{10}$
$\boxed{7} + \boxed{3} = \boxed{10}$

(b) 3, 7, 4
$\boxed{3} + \boxed{4} = \boxed{7}$
$\boxed{4} + \boxed{3} = \boxed{7}$

⑥ Use the numbers and signs to write 2 correct equations.

(a) 6, 2, =, 4, +
$4 + 2 = 6$
$2 + 4 = 6$

(b) =, 7, +, 7, 0
$7 + 0 = 7$
$0 + 7 = 7$

(c) +, 4, 9, 5, =
$4 + 5 = 9$
$5 + 4 = 9$

Note: Writing the total first is also valid, e.g. 6 = 4 + 2.

3-4 Addition with Number Bonds   39

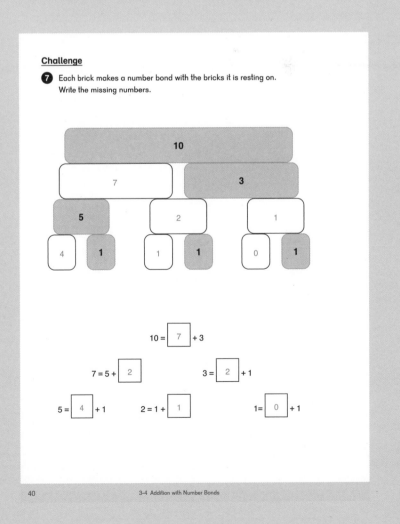

Challenge

⑦ Each brick makes a number bond with the bricks it is resting on.
Write the missing numbers.

10 = $\boxed{7}$ + 3

7 = 5 + $\boxed{2}$ 3 = $\boxed{2}$ + 1

5 = $\boxed{4}$ + 1   2 = 1 + $\boxed{1}$ 1 = $\boxed{0}$ + 1

40   3-4 Addition with Number Bonds

Exercise 5

Basics

1

| 1 | 2 | 3 | 4 | 5 | 6 | 7 | 8 | 9 | 10 |

(a) Add 0 to 7.

$7 + 0 = \boxed{7}$

(b) Add 1 to 7.

$7 + 1 = \boxed{8}$

(c) Add 2 to 7.

$7 + 2 = \boxed{9}$

(d) Add 3 to 7.

$7 + 3 = \boxed{10}$

2 (a) 2 more than 6 is __8__.

$6 + 2 = \boxed{8}$

(b) 6 more than 2 is __8__.

$2 + 6 = \boxed{8}$

(c) $6 + 2 = 2 + \boxed{6}$

Practice

3 (a) 3 more than 5 is __8__.

$5 + \boxed{3} = \boxed{8}$

(b) 8 more than 1 is __9__.

$1 + \boxed{8} = \boxed{9}$

(c) 2 added to 7 is __9__.

$7 + \boxed{2} = \boxed{9}$

3-5 Addition by Counting On   41

4 Match.

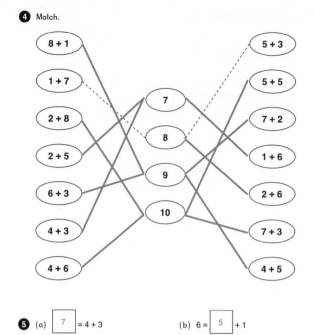

8 + 1
1 + 7
2 + 8
2 + 5
6 + 3
4 + 3
4 + 6

7
8
9
10

5 + 3
5 + 5
7 + 2
1 + 6
2 + 6
7 + 3
4 + 5

5 (a) $\boxed{7} = 4 + 3$

(b) $6 = \boxed{5} + 1$

(c) $7 + 3 = \boxed{10}$

(d) $\boxed{7} + 2 = 9$

(e) $10 = 8 + \boxed{2}$

(f) $9 + \boxed{1} = 10$

42   3-5 Addition by Counting On

Exercise 6

Basics

1

Sofia has 5 toy cars.
She buys 2 more.
How many toy cars does she have now?

$\boxed{5} \enspace \bigcirc{+} \enspace \boxed{2} = \boxed{7}$

She has __7__ toy cars now.

Practice

2

7 children had a slice of cheese pizza for lunch.
3 children had a slice of green pepper pizza for lunch.
How many slices of pizza were eaten altogether?

$\boxed{7} \enspace \bigcirc{+} \enspace \boxed{3} = \boxed{10}$

There were __10__ slices of pizza eaten altogether.

3-6 Make Addition Stories   43

3 There are 6  and 3 .
How many fish are there in all?

$\boxed{6} \enspace \bigcirc{+} \enspace \boxed{3} = \boxed{9}$

There are __9__ fish in all.

4 Dion had 4 apples.
He got 3 more.
How many apples does he have altogether?

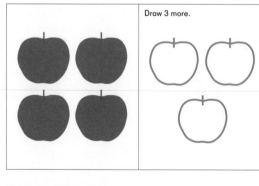

Draw 3 more.

$\boxed{4} \enspace \bigcirc{+} \enspace \boxed{3} = \boxed{7}$

He has __7__ apples altogether.

44   3-6 Make Addition Stories

Exercise 7

Basics

1

1 + 1 **2**	1 + 2 3	1 + 3 4	1 + 4 5	1 + 5 6	1 + 6 7	1 + 7 8	1 + 8 9	1 + 9 10
2 + 1 3	2 + 2 4	2 + 3 5	2 + 4 6	2 + 5 **7**	2 + 6 8	2 + 7 9	2 + 8 10	
3 + 1 4	3 + 2 5	3 + 3 6	3 + 4 7	3 + 5 8	3 + 6 9	3 + 7 10		
4 + 1 5	4 + 2 6	4 + 3 7	4 + 4 8	4 + 5 9	4 + 6 10			
5 + 1 6	5 + 2 **7**	5 + 3 8	5 + 4 9	5 + 5 10				
6 + 1 7	6 + 2 8	6 + 3 9	6 + 4 10					
7 + 1 8	7 + 2 9	7 + 3 10						
8 + 1 9	8 + 2 10							
9 + 1 10								

1	+ 4 = 5
2	+ 4 = 6
3	+ 4 = 7
4	+ 4 = 8
5	+ 4 = 9

Note: Students may see the pattern that if one part stays the same, the other part increases by the same amount as the whole.

Practice

2 Match.

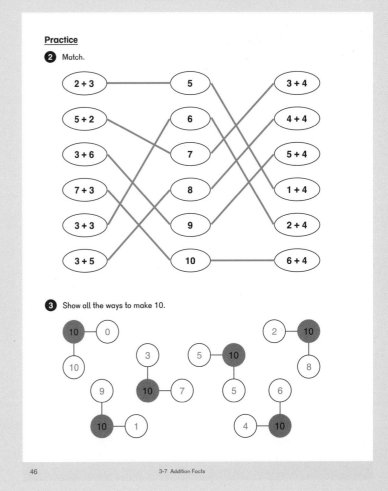

2 + 3 ── 5
5 + 2
3 + 6 ── 7
7 + 3 ── 8
3 + 3 ── 9
3 + 5 ── 10

3 + 4
4 + 4
5 + 4
1 + 4
2 + 4
6 + 4

3 Show all the ways to make 10.

10 — 0 / 10
3 / 10
9 / 10 — 7 / 10 — 1
5 — 10 / 5
2 — 10 / 8
6 / 4 — 10

4 Add.
Color the picture according to the Color Key.

Color Key				
10: Purple	**9:** Orange	**8:** Green	**7:** Pink	**6:** Brown

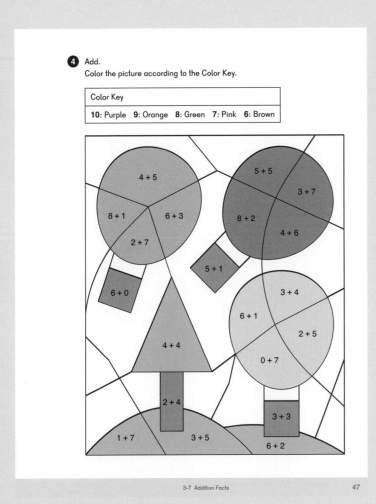

4 + 5, 5 + 5, 3 + 7, 8 + 1, 6 + 3, 8 + 2, 4 + 6, 2 + 7, 5 + 1, 6 + 0, 3 + 4, 6 + 1, 2 + 5, 0 + 7, 4 + 4, 2 + 4, 3 + 3, 1 + 7, 3 + 5, 6 + 2

5 Color the balloons in each row that match the big number.

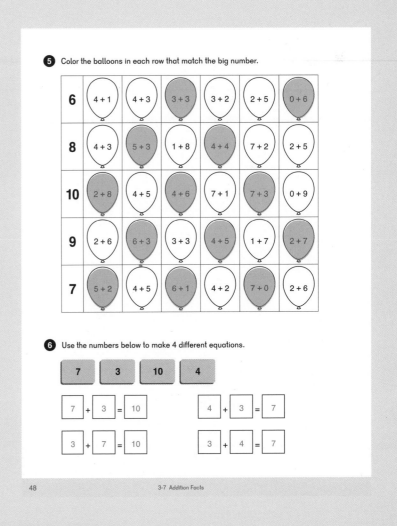

6	4 + 1	4 + 3	3 + 3	3 + 2	2 + 5	0 + 6
8	4 + 3	5 + 3	1 + 8	4 + 4	7 + 2	2 + 5
10	2 + 8	4 + 5	4 + 6	7 + 1	7 + 3	0 + 9
9	2 + 6	6 + 3	3 + 3	4 + 5	1 + 7	2 + 7
7	5 + 2	4 + 5	6 + 1	4 + 2	7 + 0	2 + 6

6 Use the numbers below to make 4 different equations.

7	3	10	4

7 + 3 = 10   4 + 3 = 7

3 + 7 = 10   3 + 4 = 7

Exercise 8

1 Circle the greatest one in each row.

(a)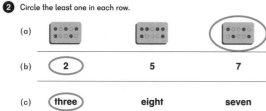

(b) **(9)**   6   8

(c) two   **(ten)**   six

(d) 5 + 4   **(7 + 3)**   2 + 6

2 Circle the least one in each row.

(a) [dot cards, third circled]

(b) **(2)**   5   7

(c) **(three)**   eight   seven

(d) 2 + 7   **(3 + 4)**   5 + 3

(e) 4 + 6   **(4 + 2)**   2 + 8

3 Put the animals in pens by drawing lines to match.

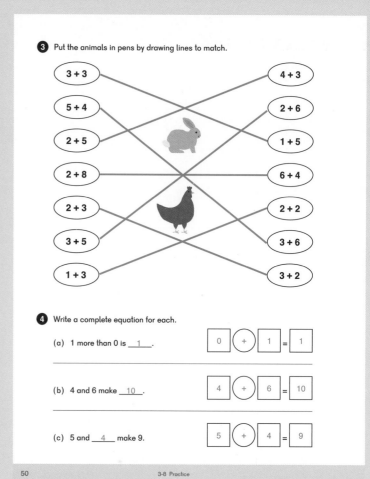

3 + 3   4 + 3
5 + 4   2 + 6
2 + 5   1 + 5
2 + 8   6 + 4
2 + 3   2 + 2
3 + 5   3 + 6
1 + 3   3 + 2

4 Write a complete equation for each.

(a) 1 more than 0 is __1__.   0 (+) 1 = 1

(b) 4 and 6 make __10__.   4 (+) 6 = 10

(c) 5 and __4__ make 9.   5 (+) 4 = 9

5

There are 3 cookies in one jar.
There are 5 cookies in the other jar.
How many cookies are there in all?

3 (+) 5 = 8

There are __8__ cookies in all.

6 There are 7 dogs playing.
3 more dogs come.
How many dogs are there in all?

7 (+) 3 = 10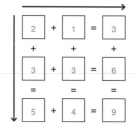

There are __10__ dogs in all.

7 Use the numbers and signs to write a correct equation for each.

(a)  4, =, 8, +, 4   _____ 4 + 4 = 8 _____

(b)  2, +, 7, =, 5   _____ 2 + 5 = 7 or 5 + 2 = 7 _____

8 (a) 5 + [5] = 10   (b) 4 + [6] = 10

(c) [8] + 2 = 10   (d) [3] + 7 = 10

Challenge

9 (a) 6 + 1 = 5 + [2]   (b) 7 + [2] = 6 + 3

(c) 7 + 3 = [9] + 1   (d) [5] + 2 = 4 + 3

10 Fill in the boxes using 1, 2, 3, 3, 4, 5, 6, 9.
Add across and down.

[2] + [1] = [3]
+ + +
[3] + [3] = [6]
= = =
[5] + [4] = [9]

Hint: If students need help, suggest they
start by trying the 9 in the lower right box
and think of two ways to make 9 with the
given numbers.

Answers may vary.

Suggested number of class periods: 11 – 12

Lesson		Page	Resources		Objectives
	Chapter Opener	p. 83	TB:	p. 55	Tell subtraction stories.
1	Subtraction as Taking Away	p. 84	TB: WB:	p. 56 p. 53	Understand subtraction as "taking away." Solve subtraction problems.
2	Subtraction as Taking Apart	p. 86	TB: WB:	p. 59 p. 55	Understand subtraction as "taking apart." Solve subtraction problems.
3	Subtraction by Counting Back	p. 88	TB: WB:	p. 61 p. 57	Subtract 1, 2, or 3 from a number within 10 by counting back.
4	Subtraction with 0	p. 90	TB: WB:	p. 63 p. 59	Understand the meaning of subtracting 0 from a number. Understand subtraction that results in a difference of 0.
5	Make Subtraction Stories	p. 92	TB: WB:	p. 65 p. 61	Tell subtraction stories given a subtraction situation.
6	Subtraction with Number Bonds	p. 95	TB: WB:	p. 68 p. 63	Use the part-whole number bond relationship to write different subtraction equations for the same situation.
7	Addition and Subtraction	p. 98	TB: WB:	p. 71 p. 65	Write addition and subtraction equations from a number bond.
8	Make Addition and Subtraction Story Problems	p. 100	TB: WB:	p. 73 p. 69	Make addition and subtraction stories.
9	Subtraction Facts	p. 102	TB: WB:	p. 75 p. 73	Learn the subtraction facts with minuends of 10 or less to mastery.
10	Practice	p. 104	TB: WB:	p. 77 p. 77	Practice the subtraction facts with minuends of 10 or less to mastery.
	Review 1	p. 106	TB: WB:	p. 80 p. 79	Review concepts from Chapter 1 through Chapter 4.
	Workbook Solutions	p. 109			

Lessons in this chapter will generally include concrete, pictorial, and abstract representations. It is important that teachers not omit the concrete portion of a lesson as many students have memorized some basic skills but have not developed a true understanding of concepts.

It is essential that students know their number bonds and addition facts to 10 fluently by this point so they can relate subtraction to addition (for example, to do 9 – 7, think of 7 and ? make 9). Students that are not fluent in addition facts should continue to practice the activities from the prior chapter.

Just as with addition, students need a large variety of examples to relate a subtraction equation to the part-part-whole representation of a number bond. In beginning subtraction, students start with the "whole." The term "bigger number" can lead to confusion later when learning the vertical algorithm for subtraction with regrouping, or when students see negative numbers.

From the whole and part, students can write the equation using a question mark to represent the missing part:

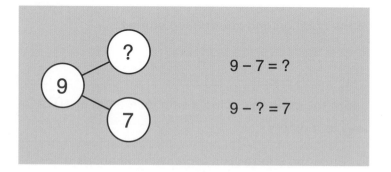

Beginning in **Lesson 7: Addition and Subtraction**, students will use the number bond to write four related addition and subtraction equations, traditionally called a "fact family."

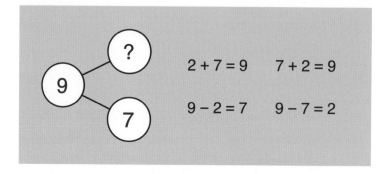

Students will be asked whether to use addition or subtraction to solve a word problem. They should not be taught to rely on key words. Guide students to determine if the problem can be solved with addition or subtraction by asking questions:

"Do we know the parts? How can we find the whole?"

"Do we know the whole and a part? How can we find another part?"

Lesson 8: Make Addition and Subtraction Story Problems and **Lesson 9: Subtraction Facts** encourage students to practice their subtraction facts to automaticity.

Multiple strategies for subtraction are taught in this chapter. These strategies are meant to enable students to find the answer to subtraction expressions. Any correct strategy is acceptable. Emphasis should be on student mastery of the facts.

Some notes on subtraction terminology:

Difference: The result of a subtraction equation is known as the difference. The difference of 5 and 3 is 2. Students are not expected to master this term; however, teachers should use it properly and suggest the word when natural to do so.

Taking away: The idea that something is removed or "goes away." Other verbs used in word problems might be "eaten," "ran away/flew away," "disappeared," "broke down," etc. In each of these situations, the initial quantity changes. These verbs can indicate the situation, but depending on which quantity is missing in the situation, the operation to find the missing number may not be subtraction.

Not every subtraction problem fits the analogy of "taking away." Here are some other ideas used as well:

Taking apart: The notion of taking apart is introduced to help students understand subtraction is not always "taking away." Taking apart means to break the whole into parts. It is used especially when we can identify two separate groups by attributes. For example, "There are 7 birds; 3 are red and 4 are blue." In taking apart, the initial quantity remains constant.

Counting Back: Students should be able to subtract without counting back. However, students will later use mental math strategies that involve counting back 1, 2, or 3. The goal here is for students to count back 1, 2, or 3 mentally, without use of manipulatives. In the **Dimensions Math®** series, counting back is suggested as a strategy only when we are subtracting 1, 2, or 3.

While we do say numbers increase and decrease, we use the phrase "counting back" rather than "counting down" because the word "down" implies movement, such as down the page. When teachers ask students to use a hundred chart to 100 to count on, students can confuse this process with the term "count down" since they are moving down the chart. Students are often confused by hundred charts that have numbers increase as they move down, while teachers say "count up."

Note: Subtraction is also used to compare numbers, that is, find the difference between two numbers. Students will learn this use of subtraction in **Dimensions Math® Textbook 1B Chapter 11: Comparing**.

Note: The following terms are educator terminology that students are not expected to learn:

Minuend: The first number in a subtraction equation. The number from which another number (the subtrahend) is to be subtracted.

Subtrahend: The number that is to be subtracted. The second number in a subtraction equation.
minuend − subtrahend = difference

Materials

- Sticky notes, 2 different colors
- Two-color counters
- Dice, 6-sided or 10-sided
- Playing cards
- Blank dice
- Dry erase sleeves
- Game markers
- Linking cubes
- Counters in a variety of shapes and colors
- Index cards or construction paper
- Sidewalk chalk
- Painter's tape
- Game board
- Whiteboards

Note: Materials for Activities will be listed in detail in each lesson.

Blackline Masters

- Number Cards
- Ten-frame Cards
- Roll and Subtract
- Flip and Count Subtraction
- Subtracting Zero Alligator Cards
- Subtraction Story Template
- Subtraction Within 10 Cover-up Cards
- Equation Symbol Cards
- Spinner

Storybooks

- Dr. Suess books
- Nonfiction nature books
- *Ten, Nine, Eight* by Molly Bang
- *Rooster's Off to See the World* by Eric Carle
- *Pete the Cat and His Four Groovy Buttons* by James Dean and Eric Litwin
- *How Many Blue Birds Flew Away?: A Counting Book with a Difference* by Paul Giganti, Jr.
- *Splash!* by Ann Jonas
- *Hot Rod Hamster* by Cynthia Lord
- *Elevator Magic* by Stuart J. Murphy
- *Ten Timid Ghosts* by Jennifer O'Connell
- *Skippyjon Jones* books by Judy Schachner

Letters Home

- Chapter 4 Letter

Chapter Opener

Objective

- Tell subtraction stories.

Have students look at the **Chapter Opener** and tell number stories about what is happening. Ask them to start with how many of something they see in all. For example, "7 dogs are playing in the park. 4 are brown. 3 are white."

Examples:

- Mei has 9 balloons. She is giving 2 balloons to Dion.
- Mei has 9 balloons. 6 balloons are red and 3 balloons are yellow.
- There are 6 children in the park. 2 of them are leaving.
- There are 9 ducks swimming in a pond. 7 are large. 2 are small.
- 7 dogs are playing in the park. 3 are large (white). 4 are small (brown).

This lesson may continue straight from **Think** to **Lesson 1: Subtraction as Taking Away**.

Activity

▲ **School Field Trip**

Have students walk around the classroom or playground and look for things that can be counted and grouped.

Examples:

- There are 6 swings. 3 have kids swinging. 3 do not have kids swinging.
- I found 10 leaves. 2 leaves are red. 8 leaves are brown.
- There are 6 boards in the hallway. 1 has math on it. 5 have reading on them.

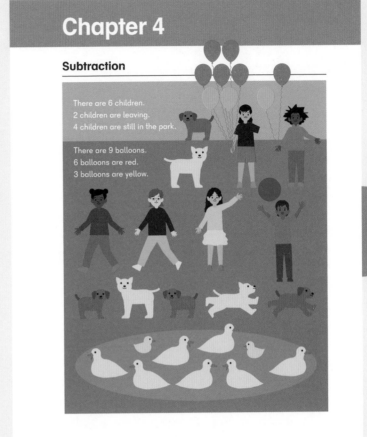

Chapter 4

Subtraction

There are 6 children.
2 children are leaving.
4 children are still in the park.

There are 9 balloons.
6 balloons are red.
3 balloons are yellow.

55

Lesson 1 Subtraction as Taking Away

Objectives

- Understand subtraction as "taking away."
- Solve subtraction problems.

Lesson Materials

- Sticky notes, 2 colors
- Counters or attribute bears
- Linking cubes, 10 per student

Think

Provide students with linking cubes and pose the **Think** problem. Ask students if they can write a number bond for the problem.

Learn

Have students show 8 cubes. Have them take 5 cubes away from their 8 and put them behind their backs. Discuss what changed in the problem: 5 ants left the log. Ensure students relate "left the log" to something "going away."

Help students with the semantics of "left." Sometimes, we refer to the ants as having left; sometimes we refer to the ants that remain, meaning that they are left (over).

Tell students that when we are taking away, we are subtracting. Just like in addition, we can use a number bond to help us make a subtraction equation:

Write "8" on one color of sticky note and put it in the whole.

Using two of the other color of sticky notes, write "5" and "?" from the number story and put them in the parts on a number bond on the board.

Review the subtraction symbol and move the sticky notes for the parts to show how they relate to the parts of a subtraction equation.

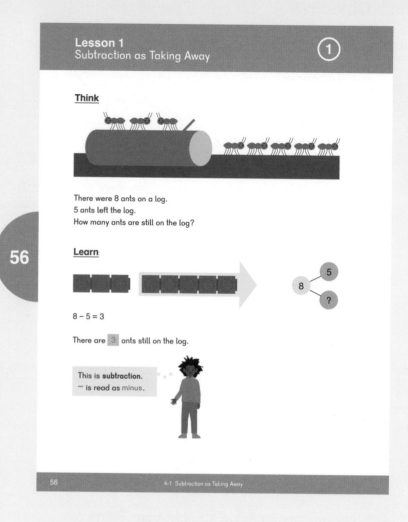

Move the sticky note from the whole in the number bond to the beginning of the subtraction equation.

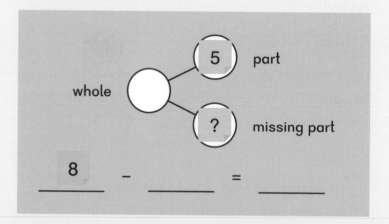

Provide students with some other examples of subtraction stories. Have students model the stories with counters, create number bonds, and write subtraction equations from the number bonds on their whiteboards. Have them use the words "minus" and "equals" when saying their equation: "7 minus 2 equals 5."

Do

Have students model the problems with cubes as needed.

Discuss other language meaning "take away," such as eaten, lost, flew away, gave, etc.

4 While both $8 - 6 = 2$ and $8 - 2 = 6$ are true equations, only one of them answers the given question. Students should practice learning not just to answer a question correctly, but to answer the correct question.

Activity

▲ Roll and Subtract

Materials: 2 dice (6 or 10-sided), Roll and Subtract (BLM)

Have students roll two dice. Then, have them write a subtraction equation created by subtracting the lesser number from the greater number to find the difference.

The Roll and Subtract (BLM) worksheet has been provided for recording purposes.

Exercise 1 • page 53

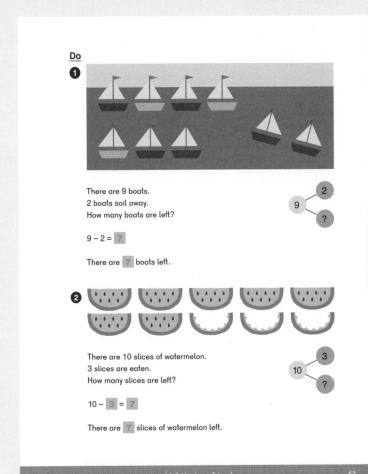

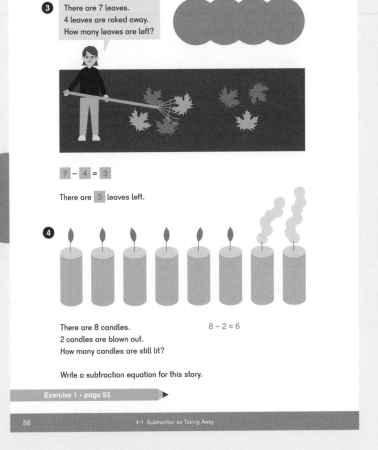

Lesson 2 Subtraction as Taking Apart

Objectives

- Understand subtraction as "taking apart."
- Solve subtraction problems.

Lesson Materials

- Two-color counters, 10 per pair of students

Think

Pose Sofia's problem with the buckets. Provide students with counters and have them share their strategies for solving the problem.

Ask students to recall the problem with the ants from the prior lesson. Ask, "How is the problem with the buckets the same as the problem with the ants? How are the two problems different?"

Students should note that no objects are being removed from the bucket problem. Discourage students from saying, "Minus means take away." Guide students to see the relationship of whole − part = part. In the **Think**, for example, this can be said as, "7 minus 3 is 4." 7 is the whole, 3 and 4 are parts.

Learn

Discuss the number bond and equations for 7 − 3 = 4.

Provide additional examples as needed and have students write the equations.

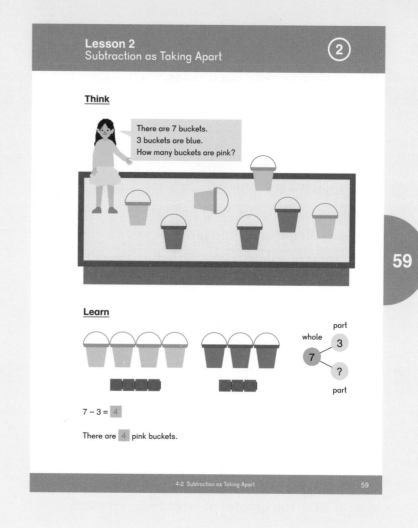

Do

To extend, have students write or illustrate word problems similar to the **Do** questions.

Activity

▲ Flip and Count Subtraction

Materials: 10 two-color counters per student, Flip and Count Subtraction (BLM) in dry erase sleeves for each student, dry erase markers

Students shake up to 10 counters and put them on Flip and Count Subtraction (BLM) in its dry erase sleeve in the correct parts.

Students fill in: whole − part = part.

Exercise 2 • page 55

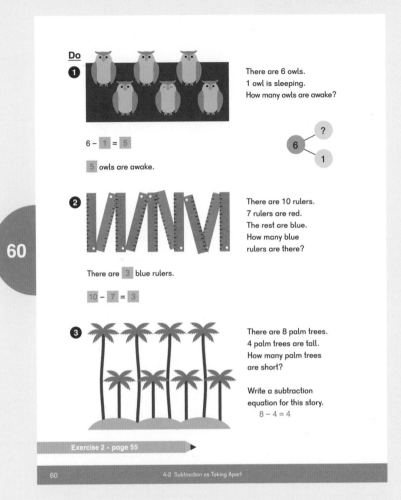

Lesson 3 Subtraction by Counting Back

Objective

- Subtract 1, 2, or 3 from a number within 10 by counting back.

Lesson Materials

- Two-color counters

Think

Have 8 students go to the front of class and ask the rest how many students there are. (8 students)

Have one student sit down and say, "7." Have another student sit down and say, "6." Have another student sit down and say, "5." Ask, "How many students sat down?" On the board, write:

$$8 - 3 = 5$$

Repeat, starting with different numbers and have students count aloud as you have students sit down. Stop after subtracting up to 3 and have students tell how many were subtracted and how many are left still standing. Write each equation on the board.

Pose Sofia's problem with the grapes. Have students model the solution with counters.

Learn

Have students look at the examples of counting back in the textbook. Then, have students do one or more of the counting back activities listed.

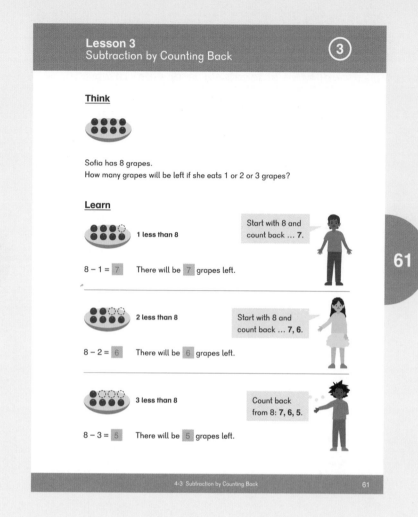

Do

Have students complete these problems on a whiteboard, or recreate ❶ — ❷ in the classroom or outside.

Students may incorrectly count back three from nine by saying, "9, 8, 7." Remind them that they are counting hops or steps moving backward, not counting the numbers visited.

Activities

▲ Counting-back Game

Materials: Game marker, blank die modified with faces: − 1, − 1, − 2, − 2, − 3, − 3

Have students create their own game board with spaces for 10 to 0. Students find another player, choose a game board and start on 10. Players take turns rolling the die and moving their marker along their own game board. First one to 0 wins.

▲ Roll and Count Back

Materials: Number Cards (BLM) 5, 6, 7, 8, 9, and 10, or regular playing cards of those numbers, blank die modified with faces: − 1, − 1, − 2, − 2, − 3, − 3

Players take turns drawing a Number Card (BLM) and rolling the die. They say the number that is 1, 2, or 3 less than the number drawn, based on the roll of the die.

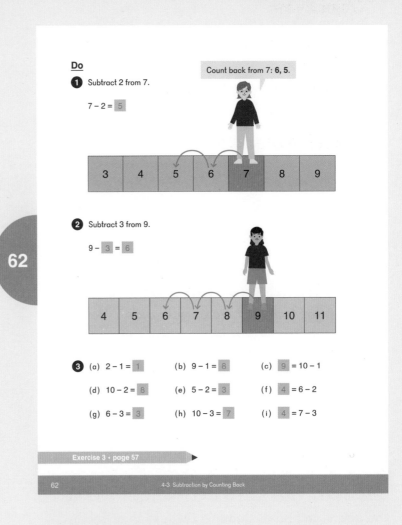

▲ Counting-back Hop

Materials: Painter's tape, counting-back dice from the **Roll and Count Back** activity

Create a number path similar to **Counting-on Hop** in **Chapter 3: Lesson 5**. Roll the counting-back dice and have the Hopper hop back from 10 to 0. When the Hopper lands on 0, players trade places.

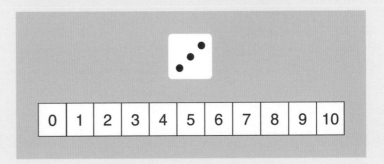

Exercise 3 • page 57

Lesson 4 Subtraction with 0

Objectives

- Understand the meaning of subtracting 0 from a number.
- Understand subtraction that results in a difference of 0.

Lesson Materials

- Counters or items from classroom

Think

Display 3 objects. Ask students, "What would happen if you took 0 of the objects away? How many would be left?"

Have students show the number bond for "3 counters take away 0" on their whiteboards.

Discuss what the equation would be and have them write an equation on their whiteboards.
Follow the same process for "3 counters take away 3 counters." Pose Alex and Emma's questions about the pears and have students discuss their solutions.

Learn

Have students discuss the equations and number bonds from **Learn**.

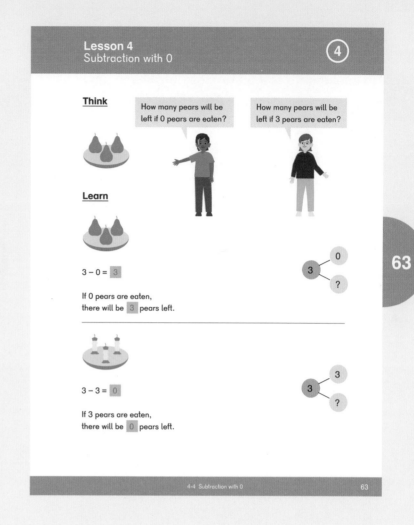

Do

3 (f) and (h) This is the first time students have seen a missing minuend. Remind students that they aren't simply subtracting numbers to find an answer; they are finding the relationship between numbers. While this is a subtraction equation, students will need to add to solve it.

To extend, ask students to create a number story that might go along with some of the equations.

Activities

▲ Alligator! Alligator! Alligator!

Materials: Subtracting Zero Alligator Cards (BLM)

Flash cards for subtraction with 0 can be used with the whole class to play a subtraction version of the **Alligator, Alligator, Alligator** game from the previous lesson.

Show Subtracting Zero Alligator Cards (BLM) to group. When you say, "Think, 2, 1, answer," they should answer chorally. When an alligator comes up, students shout, "Alligator! Alligator! Alligator!"

▲ Roll and Subtract With Zero

Materials: Roll and Subtract (BLM), 2 dice with faces: 0, 1, 2, 3, 4, 5

Two students take turns rolling two dice and creating subtraction equations on a whiteboard. The Roller rolls and writes the equation on the board. The Checker verifies the equation. For this activity, the greater number is always the whole and the lesser number is the part that is subtracted.

For assessment purposes, students can record their problems on the Roll and Subtract (BLM) worksheet.

Exercise 4 • page 59

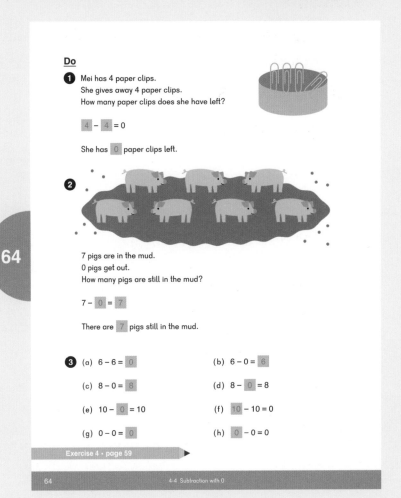

64

Do

1 Mei has 4 paper clips.
She gives away 4 paper clips.
How many paper clips does she have left?

4 – 4 = 0

She has 0 paper clips left.

2

7 pigs are in the mud.
0 pigs get out.
How many pigs are still in the mud?

7 – 0 = 7

There are 7 pigs still in the mud.

3 (a) 6 – 6 = 0   (b) 6 – 0 = 6

(c) 8 – 0 = 8   (d) 8 – 0 = 8

(e) 10 – 0 = 10   (f) 10 – 10 = 0

(g) 0 – 0 = 0   (h) 0 – 0 = 0

Exercise 4 • page 59

64   4-4 Subtraction with 0

Lesson 5 Make Subtraction Stories

Objective

- Tell subtraction stories given a subtraction situation.

Lesson Materials

- Two-color counters

Think

Have students look at the picture of the pens in **Think**. Ask students to make up a subtraction story for the pens and model the number stories with counters. Note that these stories begin with a whole and subtract one part. Encourage students to create both take away stories and take apart stories.

Write the subtraction equations as students tell the stories. Examples:

- There are 6 pens. 2 are being picked up. There are 4 pens left.
- There are 6 pens. 5 pens have caps. 1 pen does not have a cap.
- There are 6 pens. 3 are red. 3 are green.
- There are 6 pens. 2 have clips. 4 do not have clips.
- There are 6 pens. 1 has yellow on it. 5 do not have yellow.

Learn

Have students discuss Sofia's comments. Encourage students to make the connection between taking away and taking apart. For example, "If we take away the pens with the caps, the pen that is left will be the one without a cap."

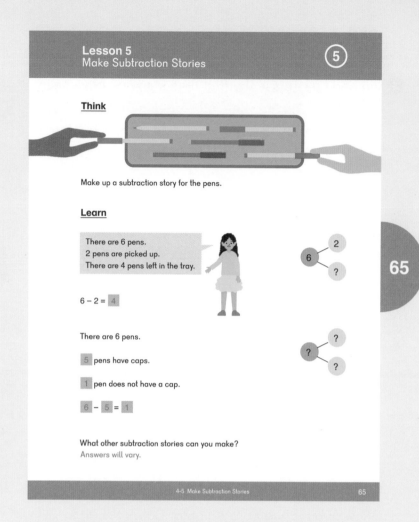

Do

1 Have students share their number stories. Guide students to identify the whole and the parts in the stories. Possible student answers:

(a) Pancakes
- There are 8 pancakes in all. 3 pancakes are in a pan. 5 pancakes are on the plate.

(b) Snails
- There are 7 snails altogether. 3 snails are blue (on the rock). 4 snails are orange (on the grass).
- 4 snails are not in their shells. 3 snails are in their shells.

(c) Blocks
- There are 10 blocks altogether. 4 blocks are red. 6 blocks are blue.
- 4 blocks are short. 6 blocks are long.
- 4 are stacked and 6 are not stacked (loose).

Activity

▲ Draw a Subtraction Story

Materials: Subtraction Story Template (BLM)

Have students create and illustrate subtraction number stories. They should also show the number bond and subtraction equation for their story. They do not need to write the story out in words.

The Subtraction Story Template (BLM) is available, or have students be creative.

Students can share their number stories.

The stories can also be collected and made into a classroom book.

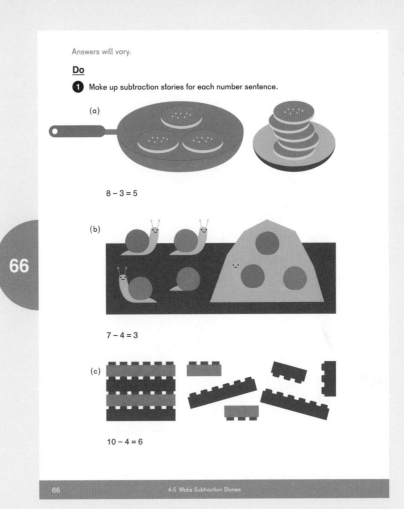

Answers will vary.

Do

1 Make up subtraction stories for each number sentence.

(a)

$8 - 3 = 5$

(b)

$7 - 4 = 3$

(c)

$10 - 4 = 6$

66   4-5 Make Subtraction Stories

66

Teacher's Guide 1A Chapter 4

2 Students should write the equations on a whiteboard. Some possible stories include:

(a) Pizza

- There are 8 pieces of pizza. 6 are on a pizza pan. 2 are not. 8 − 6 = 2
- 5 slices have pepperoni on them and 3 slices have no pepperoni. 8 − 5 = 3
- 4 slices have red sauce and 4 slices have white sauce. 8 − 4 = 4
- 2 slices have only red sauce. 6 slices have toppings. 8 − 2 = 6
- 1 slice has green peppers. 7 slices do not have green peppers. 8 − 1 = 7

(b) Dragonflies

- There are 9 dragonflies. 1 is on the flower and 8 are not. 9 − 1 = 8
- 3 have pink wings and 6 have blue wings. 9 − 3 = 6
- 2 are small and 7 are big. 9 − 2 = 7
- 5 have green bodies and 4 have orange bodies. 9 − 5 = 4

(c) Balloons

- There are 7 balloons. 3 are flying away and 4 are on the ground. 7 − 3 = 4
- 1 has green and 6 do not have green. 7 − 1 = 6
- 2 have pink baskets and 5 have purple baskets. 7 − 2 = 5

Answers will vary.

2 Make up different subtraction stories. Write an equation for each story.

8 − 1 = 7

(a) There are 8 slices of pizza. 1 slice has green peppers. 7 slices do not.

(b)

(c)

Exercise 5 • page 61

4-5 Make Subtraction Stories   67

67

Activity

▲ **Illustration Number Stories**

Materials: Storybooks or children's magazines

Almost any storybook or children's magazine lends itself to making number stories from pictures. Some suggestions:

- *Skippyjon Jones* books by Judy Schachner
- Dr. Seuss books
- *Hot Rod Hamster* by Cynthia Lord
- Nonfiction nature books

◀ **Exercise 5 • page 61**

Lesson 6 Subtraction with Number Bonds

Objective

- Use the part-whole number bond relationship to write different subtraction equations for the same situation.

Lesson Materials

- Linking cubes, 10 each of 2 different colors per student
- Different colored shaped counters

Think

Give each student the linking cubes and have them show 5 linking cubes of one color and 3 of another color.

Have students tell the number bonds and come up with subtraction stories for the goats.

Students should find that when given the whole of 8 they can create two subtraction equations:

8 − 5 = 3 and 8 − 3 = 5
or
3 = 8 − 5 and 5 = 8 − 3

Have students share their equations and write them on the board. Remind them to refer to 8 as the whole, rather than the greatest (or biggest) number.

Learn

Provide students with other number bonds with missing parts on the board and have them create the equations on their whiteboards. They can use the linking cubes to model and create a number story.

Discuss the equations in **Learn**.

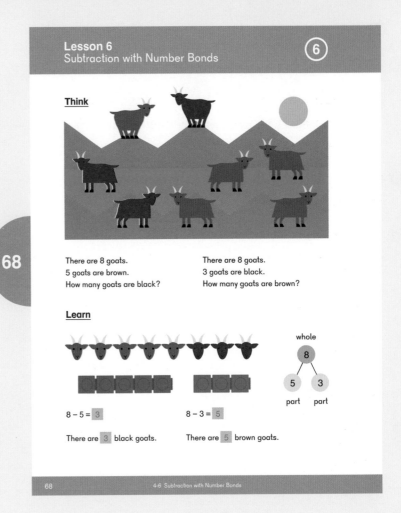

Do

❸ Possible student responses:

- There are 6 chickens. 2 are roosters. 4 are chicks.
- There are 6 chickens. 4 are small. 2 are big.
- There are 6 chickens. 2 are brown. 4 are yellow.

❹ — ❺ Have each student solve the problems on a whiteboard and check for understanding.

❺ (b), (c), and (e) This is the first time students have related a missing minuend problem to a number bond. While this is a subtraction equation, students will need to add to solve it.

The practice activity below can be used as an extension.

Activities

▲ Make a Match

Materials: Subtraction within 10 Cover-up Cards (BLM), 1 set per pair of students

This game is similar to **Chapter 3: Lesson 4**, on page 60 of this Teacher's Guide. Have students lay the set of Subtraction within 10 Cover-up Cards (BLM) facedown and have them take turns turning over two cards. If the two cards have the same answer, the student keeps those cards.

For example, if Player 1 draws:

| 7 – 3 = 4 | 7 – 5 = 2 |

She does not have a match, so she will return the cards to the pile and play will continue.

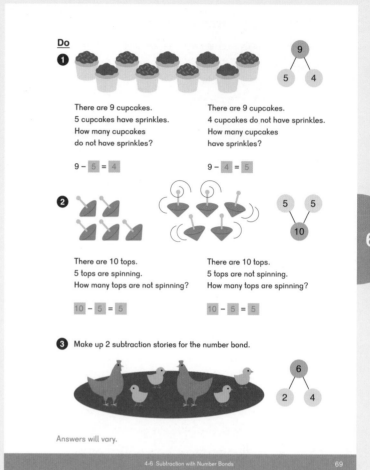

If Player 2 draws:

| 5 – 3 = 2 | 6 – 4 = 2 |

She has a match and keeps the cards.

The player with the most cards at the end is the winner.

▲ Subtraction Cover-up

Materials: Subtraction Within 10 Cover-up Cards (BLM)

Shuffle Subtraction Within 10 Cover-up Cards (BLM) and show them to students using your finger or a strip of paper to cover one of the parts or the whole. Have students tell you the number that is covered up.

★ Make Subtraction Equations

Materials: Several sets of Number Cards (BLM) 0 to 10, Equation Symbol Cards (BLM)

Ask students, "Using all these cards, can you make two subtraction equations?"

Students may need to try different equations to solve the problem correctly:

$9 - 2 = 7$ or $9 - 7 = 2$

$6 - 5 = 1$ or $6 - 1 = 5$

For more challenge, add the cards 3, 4, 7, −, = (or 9, 3, 6, −, =, or create your own) and have students make three equations.

Students can record or write number stories to go with the equations.

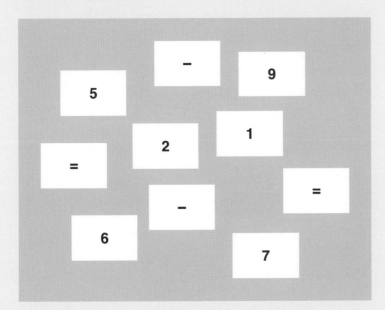

Exercise 6 • page 63

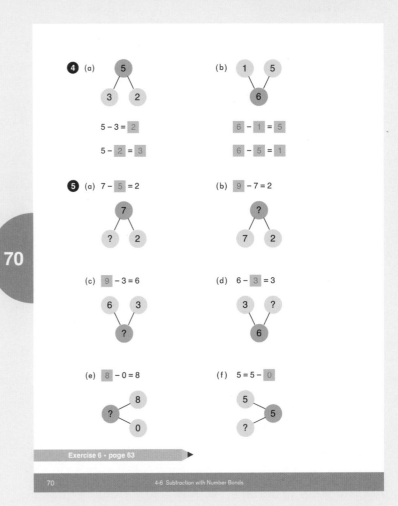

4 (a)

$5 - 3 = 2$
$5 - 2 = 3$

(b)

$6 - 1 = 5$
$6 - 5 = 1$

5 (a) $7 - 5 = 2$

(b) $9 - 7 = 2$

(c) $9 - 3 = 6$

(d) $6 - 3 = 3$

(e) $8 - 0 = 8$

(f) $5 = 5 - 0$

Exercise 6 • page 63

70   4-6 Subtraction with Number Bonds

70

Lesson 7 Addition and Subtraction

Objective

- Write addition and subtraction equations from a number bond.

Lesson Materials

- Two-color counters or 2 sets of different shaped counters, 10 per pair of students

Think

Pose the problem with the sharpeners. Have students create a number story and show the number bond on their whiteboards. Examples:

- 3 red crayons and 4 blue crayons
- 3 airplane counters and 4 truck counters

Ask students to show an addition equation for the number bond. Most will probably write:

$3 + 4 = 7$ or $4 + 3 = 7$

Ask students to come up with two subtraction equations.

$7 - 3 = 4$ or $7 - 4 = 3$

You could also include:

$7 = 4 + 3$ $7 = 3 + 4$
$4 = 7 - 3$ $3 = 7 - 4$

Note: These equations are traditionally taught as a "fact family." Here, we use this idea to emphasize all the various ways the two parts and the whole are related.

Discuss students' equations and solutions to the problems.

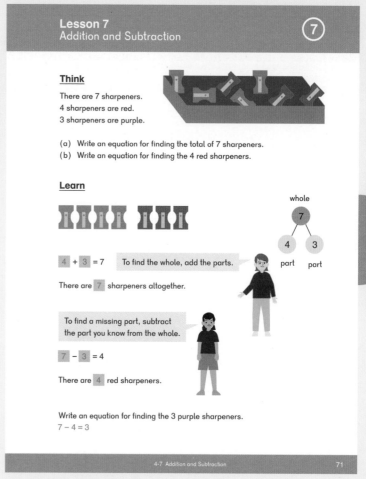

Think

There are 7 sharpeners.
4 sharpeners are red.
3 sharpeners are purple.

(a) Write an equation for finding the total of 7 sharpeners.
(b) Write an equation for finding the 4 red sharpeners.

Learn

whole

7

4   3

part   part

$4 + 3 = 7$ To find the whole, add the parts.

There are 7 sharpeners altogether.

To find a missing part, subtract the part you know from the whole.

$7 - 3 = 4$

There are 4 red sharpeners.

Write an equation for finding the 3 purple sharpeners.
$7 - 4 = 3$

4-7 Addition and Subtraction   71

Learn

Discuss Emma's and Mei's comments. Have students write the equations on their whiteboards.

Students can work in pairs to practice writing four equations. One student can create a number story with counters. The partner writes the four addition and subtraction equations on a whiteboard.

Do

② (c) Discuss with students why there aren't four equations for the number bond.

Activities

▲ Groups

The teacher (or Caller) calls out a number, and the students have 10 seconds to get themselves into groups of that size. It might be impossible for everyone to get in a group every time, but each new number gives everyone another chance.

To begin the activity, call out single numbers. Once students get the idea, call out addition or subtraction problems such as, "Groups of 7 − 4."

▲ Stand Up/Sit Down

The teacher (or Caller) gives an equation with a missing number. If the missing number is 10, students stand up.

If the missing number is anything other than 10, students sit down.

Use expressions like "7 + 3" and "9 − 2" to have students practice their facts.

Exercise 7 • page 65

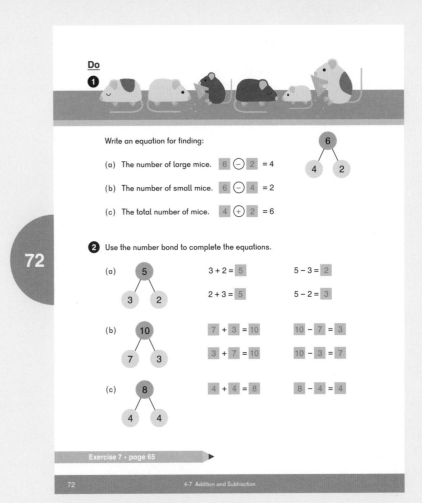

72

Lesson 8 Make Addition and Subtraction Story Problems

Objective

- Make addition and subtraction stories.

Think

Pose the problems with the orange slices. Have students write equations on whiteboards for the friends' questions:

- How many orange slices are there altogether?
- How many orange slices are small?

Discuss students' equations and solutions to the problems.

Learn

Have students look at page 73 and discuss the number bonds and equations for the two questions.

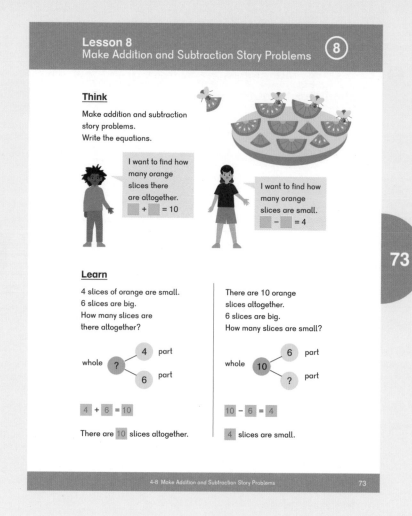

Do

1 — **2** Finding both addition and subtraction stories from a picture can be difficult. Some suggestions are below:

1 (a) Eggs

- 1 egg is broken. 5 eggs are not broken. There are 6 eggs in all. 1 + 5 = 6
- 2 eggs are in a carton. 4 eggs are on a plate. There are 6 eggs in all. 2 + 4 = 6

1 (b) Eggs

- There were 6 eggs. 1 egg broke. There are 5 whole eggs left. 6 − 1 = 5
- There were 6 eggs in a carton. 4 eggs were moved to a plate. There are 2 eggs left in the carton. 6 − 2 = 4

2 (a) Balloons

- There are 3 blue and 5 red balloons. There are 8 balloons in all. 3 + 5 = 8
- 2 balloons have smiley faces and 6 don't have smiley faces. 2 + 6 = 8
- There are 3 floating away and 5 in Sofia's hands. 3 + 5 = 8

2 (b) Balloons

- There are 8 balloons. 3 are blue. How many are not blue? 8 − 3 = 5
- There are 8 balloons. 2 of them have smiley faces, how many do not? 8 − 2 = 6
- There are 8 balloons. Sofia is holding 5. How many are floating away? 8 − 5 = 3

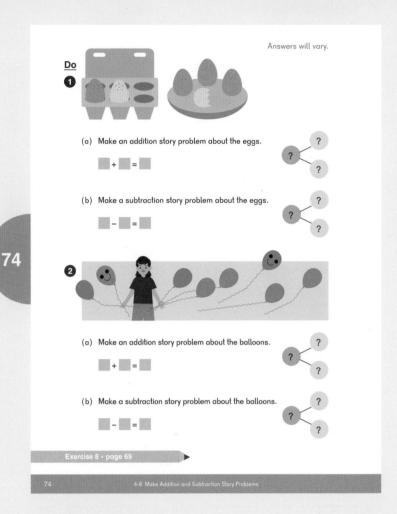

Activity

▲ Draw a Story

As in **Lesson 5: Make Subtraction Stories** of this chapter, have students create and illustrate number stories on a blank sheet of paper. Make sure they include stories that have both addition and subtraction components. They can show the number bond and equations for their story. They do not need to write the story out in words. Students can share their number stories.

Exercise 8 · page 69

Lesson 9 Subtraction Facts

Objective

- Learn the subtraction facts with minuends of 10 or less to mastery.

Lesson Materials

- Index cards, 50 per student, or construction paper
- Number Cards (BLM) 1 to 10, 1 set per student
- Spinner (BLM) or 10-sided die
- Paper clip for Spinner
- Ten-frame Cards (BLM) 1 to 10

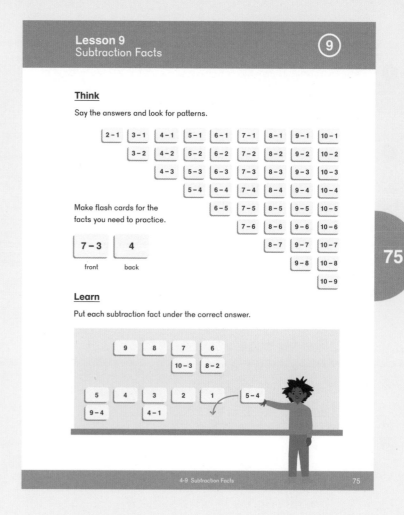

Think

Have students find the answers and look for patterns in the fact cards shown. There are many different patterns students may notice.

To extend, have students find different ways of organizing their cards based on different patterns. For example, they may try lining up the cards by the numbers being subtracted: 10 − 5, 9 − 5, 8 − 5, 7 − 5, ...

Learn

Provide students with index cards and have them create their own Math Fact Flash Cards for future practice and games. They can lay them out in a similar pattern while looking for patterns.

Students can also fold construction paper into eight equal parts and cut out their own flash cards.

Have students lay out Number Cards (BLM) 1 to 10 and sort their flash cards under each number.

Do

1 To extend, teachers can also show students Ten-frame Cards (BLM) and ask them, "What is 2 less? What is 3 less?"

2 Students play in pairs, using Spinner (BLM).

Activity

▲ Math Fact Jump

Materials: Math Fact Flash Cards from this lesson, sidewalk chalk or painter's tape

Create two grids as shown below, using either chalk outside or painter's tape inside.

One student is the Caller. Two students are the Jumpers and stand on their Home square. The Caller says a math fact. Jumpers must jump on the answer to the fact.

The first Jumper who misses the correct square becomes the next Caller.

0	9	5
6	7	3
1	4	2
8	10	Home

0	9	5
6	7	3
1	4	2
8	10	Home

Exercise 9 • page 73

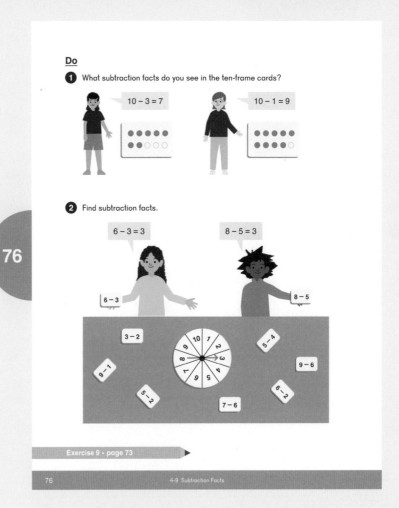

Lesson 10 Practice

Objective

- Practice the subtraction facts with minuends of 10 or less to mastery.

Have students play games from Chapter 4 to practice.

Do

2 Stories for 10 − 1 = 9:

- There are 10 bats in all. 1 is brown and 9 are gray. 10 − 1 = 9
- 2 have pink feet. 8 have gray feet. 10 − 2 = 8
- There are 3 little bats and 7 big bats. 10 − 3 = 7
- I see 4 bats flying and 6 not flying. 10 − 4 = 6
- There are 5 with fangs and 5 without fangs. 10 − 5 = 5

Activities

▲ **Takeover!**

Materials: A game board from the classroom, Math Fact Flash Cards created in **Lesson 9: Subtraction Facts**

Most board games require a roll of the dice to determine how many squares to move. In Takeover!, players use a deck of Subtraction Fact Cards to move.

On each turn, the player draws a flash card and figures out the answer. For example, if a player draws 7 − 2, he moves forward 5 spaces.

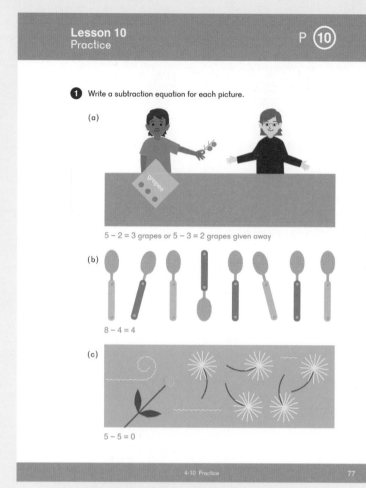

Lesson 10 Practice P 10

1 Write a subtraction equation for each picture.

(a)

5 − 2 = 3 grapes or 5 − 3 = 2 grapes given away

(b)

8 − 4 = 4

(c)

5 − 5 = 0

77

4-10 Practice   77

Answers will vary.

2 Make up subtraction stories about the bats. Write an equation for each.

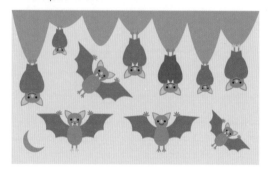

3 Find the missing number in the number bond. Use the number bond to complete the equations.

4 + 6 = 10   10 − 4 = 6

6 + 4 = 10   10 − 6 = 4

4 (a) 10 − 9 = 1   (b) 8 − 0 = 8   (c) 9 − 6 = 3

(d) 9 − 5 = 4   (e) 7 − 4 = 3   (f) 5 − 5 = 0

78

78   4-10 Practice

▲ Subtraction Face-off

Materials: Math Fact Flash Cards created in **Lesson 9: Subtraction Facts**

Each player turns two cards faceup, reads the equation, and supplies the answer. The player with the greater difference collects all four cards.

For example, Player 1 draws a 5 and a 4, and says, "5 − 4 = 1." Player 2 draws a 7 and a 2, and says, "7 − 2 = 5." Player 2's result is greater so she wins the round.

If players have the same answer, then they have to face off. At this point, you'll reverse the math operation and do an addition problem. The player with the greatest sum wins all cards.

Brain Works

★ What Numbers Work?

Have students find numbers that will complete these equations. Ask students how they know if their equations are "true."

$4 + 4 = $ _____ − _____

$3 + 1 = $ _____ − _____

$9 − 2 = $ _____ + _____

$7 − 3 = $ _____ + _____

Exercise 10 • page 77

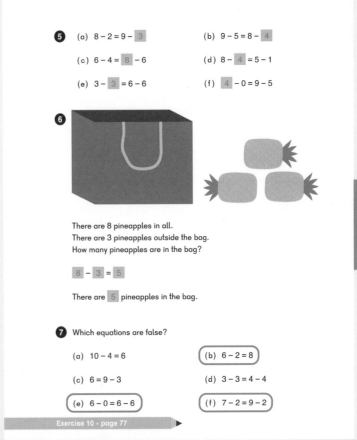

⑤ (a) $8 - 2 = 9 - \boxed{3}$ (b) $9 - 5 = 8 - \boxed{4}$

(c) $6 - 4 = \boxed{8} - 6$ (d) $8 - \boxed{4} = 5 - 1$

(e) $3 - \boxed{3} = 6 - 6$ (f) $\boxed{4} - 0 = 9 - 5$

⑥

There are 8 pineapples in all.
There are 3 pineapples outside the bag.
How many pineapples are in the bag?

$\boxed{8} - \boxed{3} = \boxed{5}$

There are $\boxed{5}$ pineapples in the bag.

⑦ Which equations are false?

(a) $10 - 4 = 6$ (b) $6 - 2 = 8$

(c) $6 = 9 - 3$ (d) $3 - 3 = 4 - 4$

(e) $6 - 0 = 6 - 6$ (f) $7 - 2 = 9 - 2$

Exercise 10 • page 77

4-10 Practice   79

79

Review 1

Objective

- Review concepts from Chapter 1 through Chapter 4.

Reviews provide cumulative practice with skills learned throughout the first four chapters. Activities and games can be used to reinforce basic skills. By the end of this chapter, students should be fluent with number combinations to 10.

Do

4 Extend this problem by providing mixed addition and subtraction fact cards. Students can put them in order independently or in small groups.

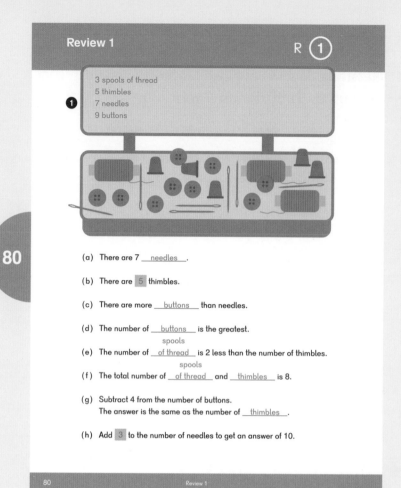

Review 1   R ①

1
- 3 spools of thread
- 5 thimbles
- 7 needles
- 9 buttons

(a) There are 7 __needles__ .

(b) There are 5 thimbles.

(c) There are more __buttons__ than needles.

(d) The number of __buttons__ is the greatest.

(e) The number of __spools of thread__ is 2 less than the number of thimbles.

(f) The total number of __spools of thread__ and __thimbles__ is 8.

(g) Subtract 4 from the number of buttons.
The answer is the same as the number of __thimbles__ .

(h) Add 3 to the number of needles to get an answer of 10.

80

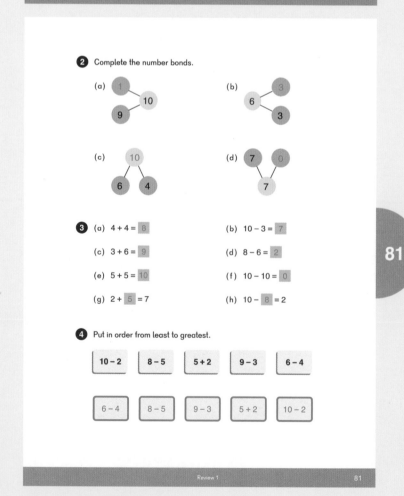

2 Complete the number bonds.

(a) 1, 9 → 10

(b) 3, 3 → 6

(c) 10 → 6, 4

(d) 7, 0 → 7

3 (a) 4 + 4 = 8   (b) 10 − 3 = 7

(c) 3 + 6 = 9   (d) 8 − 6 = 2

(e) 5 + 5 = 10   (f) 10 − 10 = 0

(g) 2 + 5 = 7   (h) 10 − 8 = 2

4 Put in order from least to greatest.

| 10 − 2 | 8 − 5 | 5 + 2 | 9 − 3 | 6 − 4 |

| 6 − 4 | 8 − 5 | 9 − 3 | 5 + 2 | 10 − 2 |

81

Brain Works

★ Number Code

0 = −

1 = *

2 = * −

3 = * *

4 = * − −

5 = * − *

6 = * * −

7 = * * *

8 = * − − −

9 = ?

10 = ?

9 = * − − *

10 = * − * −

Exercise 11 • page 79

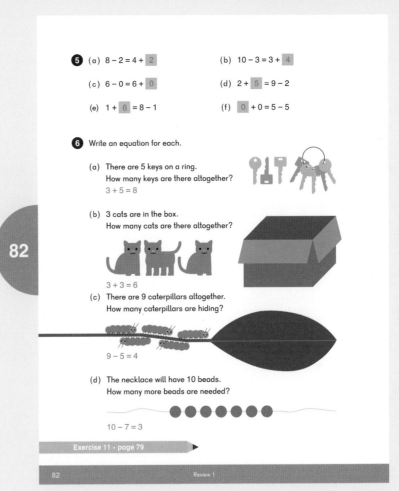

5 (a) $8 - 2 = 4 + \boxed{2}$ (b) $10 - 3 = 3 + \boxed{4}$

(c) $6 - 0 = 6 + \boxed{0}$ (d) $2 + \boxed{5} = 9 - 2$

(e) $1 + \boxed{6} = 8 - 1$ (f) $\boxed{0} + 0 = 5 - 5$

6 Write an equation for each.

(a) There are 5 keys on a ring.
How many keys are there altogether?
$3 + 5 = 8$

(b) 3 cats are in the box.
How many cats are there altogether?
$3 + 3 = 6$

(c) There are 9 caterpillars altogether.
How many caterpillars are hiding?
$9 - 5 = 4$

(d) The necklace will have 10 beads.
How many more beads are needed?
$10 - 7 = 3$

Exercise 11 • page 79

82   Review 1

Notes

Chapter 4 Subtraction

Exercise 1

Basics

1

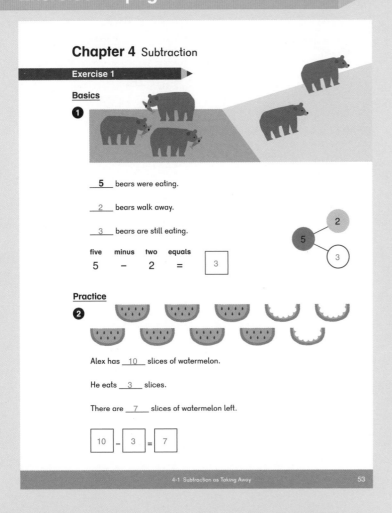

_**5**_ bears were eating.

_**2**_ bears walk away.

_**3**_ bears are still eating.

five	minus	two	equals	
5	–	2	=	3

Practice

2

Alex has _**10**_ slices of watermelon.

He eats _**3**_ slices.

There are _**7**_ slices of watermelon left.

10 – 3 = 7

3 (a)

8 – 5 = 3

(b)
6 – 2 = 4

(c)
9 – 4 = 5

(d)
7 – 3 = 4

(e)
10 – 5 = 5

Exercise 2

Basics

1

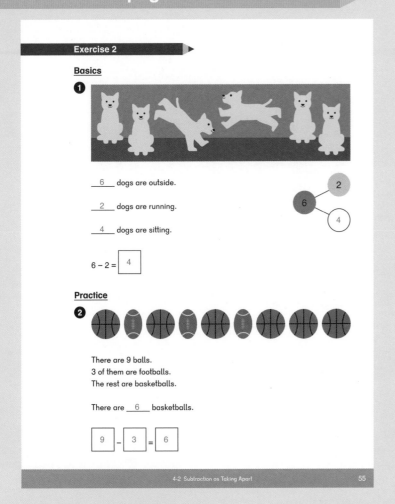

_**6**_ dogs are outside.

_**2**_ dogs are running.

_**4**_ dogs are sitting.

6 – 2 = 4

Practice

2

There are 9 balls.
3 of them are footballs.
The rest are basketballs.

There are _**6**_ basketballs.

9 – 3 = 6

3 (a)
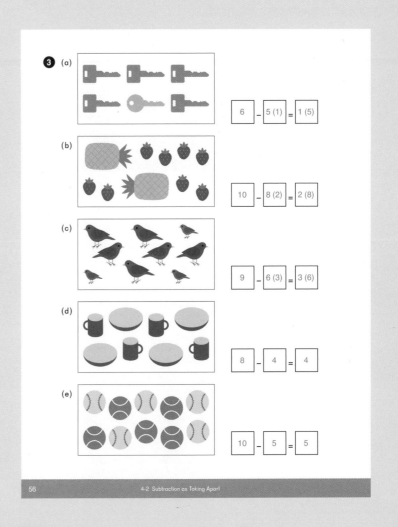
6 – 5 (1) = 1 (5)

(b)
10 – 8 (2) = 2 (8)

(c)
9 – 6 (3) = 3 (6)

(d)
8 – 4 = 4

(e)
10 – 5 = 5

Exercise 3

Basics

1

10	9	8	7	6	5	4	3	2	1

(a) 1 less than 10.

$10 - 1 = \boxed{9}$

(b) Subtract 1 from 6.

$6 - 1 = \boxed{5}$

(c) 2 less than 10.

$10 - 2 = \boxed{8}$

(d) Subtract 2 from 6.

$6 - 2 = \boxed{4}$

(e) Subtract 3 from 10.

$10 - 3 = \boxed{7}$

(f) 3 less than 6.

$6 - 3 = \boxed{3}$

Practice

2 (a)

1 2 3 4 5 6 7

$7 - 2 = \boxed{5}$

(b)

1 2 3 4 5 6 7 8

$8 - 3 = \boxed{5}$

---

3 (a) $7 \xrightarrow{-1} \boxed{6}$ (b) $9 \xrightarrow{-1} \boxed{8}$

(c) $8 \xrightarrow{-2} \boxed{6}$ (d) $10 \xrightarrow{-2} \boxed{8}$

(e) $9 \xrightarrow{-2} \boxed{7}$ (f) $7 \xrightarrow{-3} \boxed{4}$

(g) $5 \xrightarrow{-3} \boxed{2}$ (h) $6 \xrightarrow{-2} \boxed{4}$

4 (a) $10 - 1 = \boxed{9}$ (b) $\boxed{10} - 2 = 8$

(c) $\boxed{7} = 10 - 3$ (d) $9 - \boxed{2} = 7$

(e) $7 - \boxed{3} = 4$ (f) $\boxed{8} - 2 = 6$

Challenge

5 Subtract by counting on from the part to the whole.

$9 - 7 = ?$

6	7	8	9	10

$9 - 7 = 2$

(a) $10 - 9 = \boxed{1}$ (b) $6 - 4 = \boxed{2}$

(c) $7 - 5 = \boxed{2}$ (d) $8 - 5 = \boxed{3}$

Exercise 4

Basics

1

5 owls were sleeping.
Then 5 owls woke up.
How many owls are still sleeping?

$5 - 5 = \boxed{0}$

<u>0</u> owls are still sleeping.

2

Mei has 6 stuffed bears.
She gives 0 of them away.
How many bears are left?

$\boxed{6} - \boxed{0} = \boxed{6}$

<u>6</u> bears are left.

Practice

3 Write a correct subtraction equation for each.

(a) 5,   0,   =,   5,   –

     $5 - 5 = 0$ or $5 - 0 = 5$

(b) =,   7,   –,   9,   2

     $9 - 2 = 7$ or $9 - 7 = 2$

(c) –,   5,   3,   8,   =

     $8 - 3 = 5$ or $8 - 5 = 3$

4 (a) 9 less than 9 is <u>0</u> .

$\boxed{9} - \boxed{9} = \boxed{0}$

(b) 0 less than 7 is <u>7</u> .

$\boxed{7} - \boxed{0} = \boxed{7}$

(c) 3 subtracted from 3 is <u>0</u> .

$\boxed{3} - \boxed{3} = \boxed{0}$

Challenge

5 (a) 2 less than $6 - 2$ is <u>2</u> .

(b) 3 less than $10 - 3$ is <u>4</u> .

(c) 1 less than $7 - 6$ is <u>0</u> .

(d) <u>1</u> less than $9 - 1$ is 7.

(e) <u>2</u> less than $8 - 6$ is 0.

Exercise 5

Basics

1

There are 7 cows.

(a) 4 cows have spots.
The rest have no spots.
How many cows do not have spots?

___3___ cows have no spots.

$\boxed{7} \,\bigcirc\!\!-\, \boxed{4} = \boxed{3}$

(b) 2 cows have a bell around their neck.
How many cows have no bell?

___5___ cows have no bell.

$\boxed{7} \,\bigcirc\!\!-\, \boxed{2} = \boxed{5}$

(c) 1 cow has horns.
How many cows have no horns?

___6___ cows have no horns.

$\boxed{7} \,\bigcirc\!\!-\, \boxed{1} = \boxed{6}$

Practice

2 (a)

I have 9 bats in a bin.
I take out 6 bats.

$9 - 6 = \boxed{3}$

___3___ bats are still in the bin.

(b)

I have 8 sandwiches in a bag.
I put 2 on a plate.

$8 - 2 = \boxed{6}$

___6___ sandwiches are still in the bag.

(c)

I have 10 shells in a bucket.
I take out 5 shells.

$10 - 5 = \boxed{5}$

___5___ shells are still in the bucket.

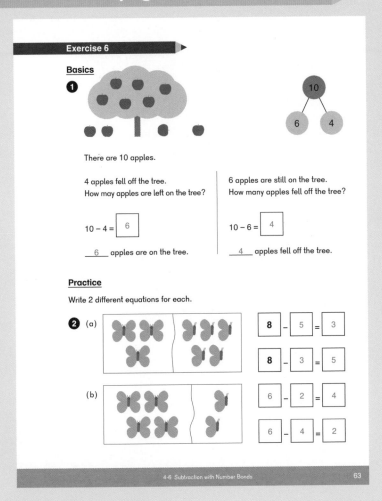

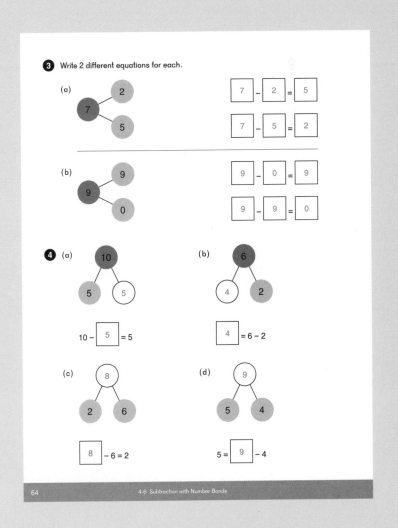

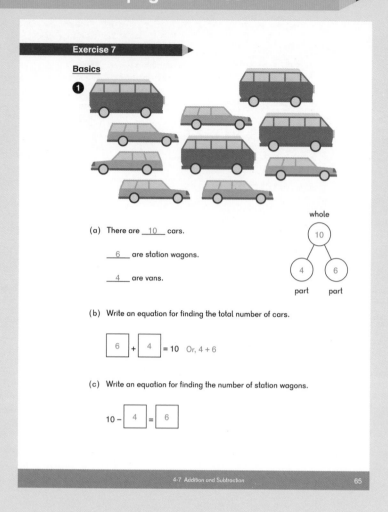

Exercise 7

Basics

1

(a) There are __10__ cars.

__6__ are station wagons.

__4__ are vans.

whole

10

4   6

part   part

(b) Write an equation for finding the total number of cars.

$6 + 4 = 10$ Or, $4 + 6$

(c) Write an equation for finding the number of station wagons.

$10 - 4 = 6$

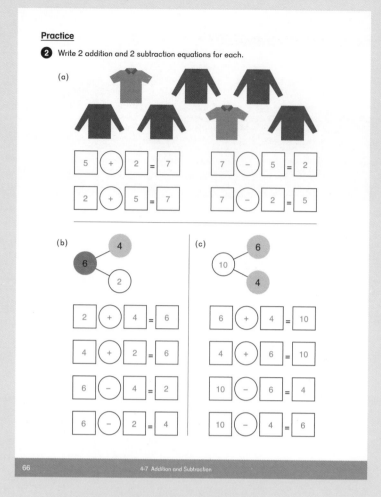

Practice

2 Write 2 addition and 2 subtraction equations for each.

(a)

$5 + 2 = 7$ $7 - 5 = 2$

$2 + 5 = 7$ $7 - 2 = 5$

(b)

4

6

2

$2 + 4 = 6$

$4 + 2 = 6$

$6 - 4 = 2$

$6 - 2 = 4$

(c)

6

10

4

$6 + 4 = 10$

$4 + 6 = 10$

$10 - 6 = 4$

$10 - 4 = 6$

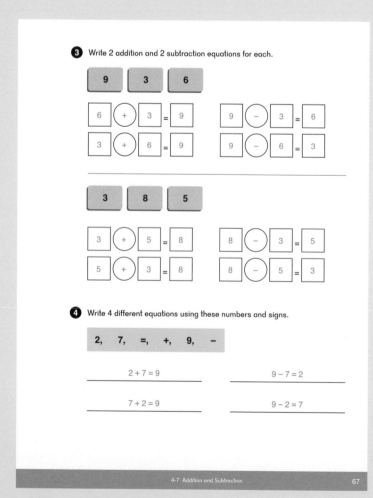

3 Write 2 addition and 2 subtraction equations for each.

| 9 | 3 | 6 |

$6 + 3 = 9$ $9 - 3 = 6$

$3 + 6 = 9$ $9 - 6 = 3$

| 3 | 8 | 5 |

$3 + 5 = 8$ $8 - 3 = 5$

$5 + 3 = 8$ $8 - 5 = 3$

4 Write 4 different equations using these numbers and signs.

| 2, | 7, | =, | +, | 9, | – |

$2 + 7 = 9$ $9 - 7 = 2$

$7 + 2 = 9$ $9 - 2 = 7$

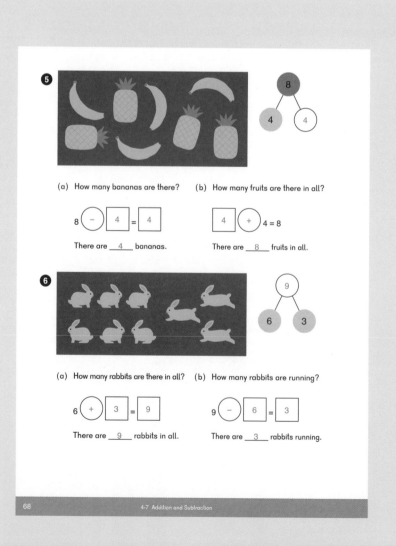

5

8

4   4

(a) How many bananas are there?

$8 - 4 = 4$

There are __4__ bananas.

(b) How many fruits are there in all?

$4 + 4 = 8$

There are __8__ fruits in all.

6

9

6   3

(a) How many rabbits are there in all?

$6 + 3 = 9$

There are __9__ rabbits in all.

(b) How many rabbits are running?

$9 - 6 = 3$

There are __3__ rabbits running.

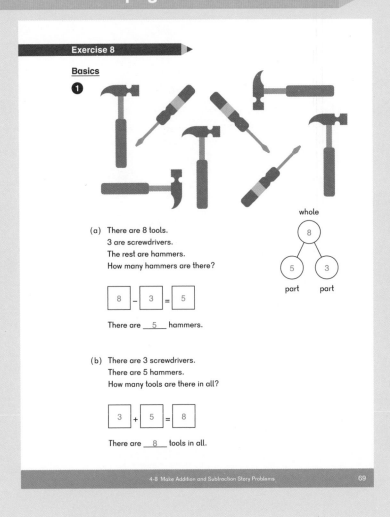

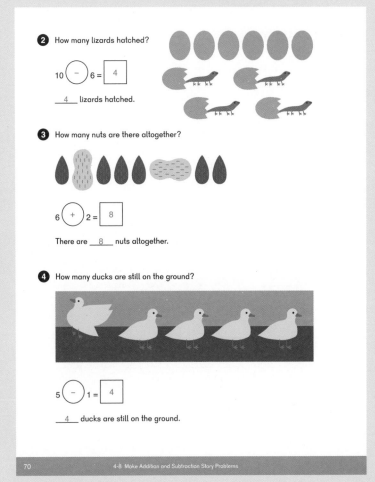

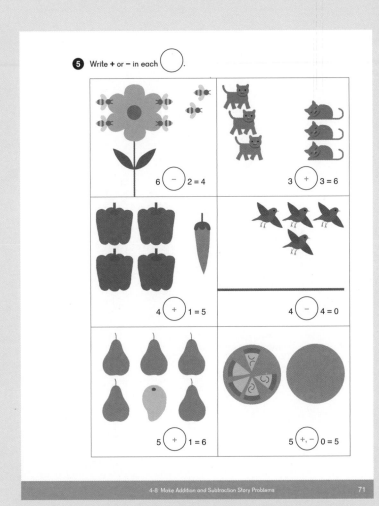

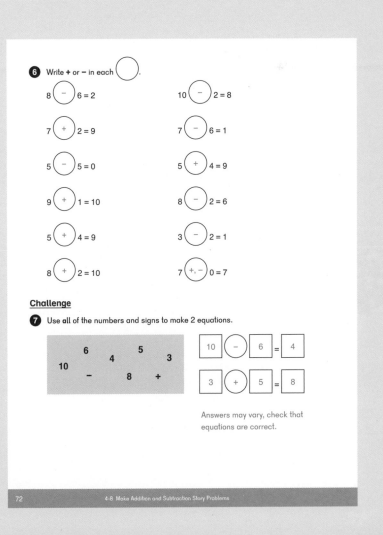

Exercise 9

Basics

1

Note: Students may see the pattern that if the whole stays the same, when one part decreases, the other part must increase by the same amount.

1 – 1 0									
2 – 1 1	2 – 2 0								
3 – 1 2	3 – 2 1	3 – 3 0							
4 – 1 3	4 – 2 2	4 – 3 1	4 – 4 0						
5 – 1 4	5 – 2 3	5 – 3 2	5 – 4 1	5 – 5 0					
6 – 1 5	6 – 2 4	6 – 3 3	6 – 4 2	6 – 5 1	6 – 6 0				
7 – 1 6	7 – 2 5	7 – 3 4	7 – 4 3	7 – 5 2	7 – 6 1	7 – 7 0			
8 – 1 7	8 – 2 6	8 – 3 5	8 – 4 4	8 – 5 3	8 – 6 2	8 – 7 1	8 – 8 0		
9 – 1 8	9 – 2 7	9 – 3 6	9 – 4 5	9 – 5 4	9 – 6 3	9 – 7 2	9 – 8 1	9 – 9 0	
10 – 1 9	10 – 2 8	10 – 3 **7**	10 – 4 6	10 – 5 5	10 – 6 4	10 – 7 3	10 – 8 2	10 – 9 1	10 – 10 0

9 – [8] = 1

9 – [7] = 2

9 – [6] = 3

9 – [5] = 4

9 – [4] = 5

Practice

2 Color the hearts in each row that match the big number.

5	5 – 5	3 – 2	7 – 2	10 – 5	7 – 1	9 – 5
4	10 – 6	9 – 3	8 – 4	5 – 1	10 – 4	6 – 2
2	10 – 7	9 – 7	4 – 1	8 – 2	6 – 4	1 – 1
3	9 – 6	5 – 3	10 – 7	8 – 5	2 – 1	7 – 4
1	8 – 6	9 – 8	5 – 2	7 – 5	4 – 3	6 – 1
6	9 – 3	4 – 2	6 – 0	8 – 2	5 – 1	7 – 1

3 (a) 8 – 4 = [4]   (b) 7 – 4 = [3]

(c) 9 – [6] = 3   (d) 6 – [4] = 2

(e) [10] – 2 = 8   (f) 8 – [8] = 0

---

4 Subtract.
Color the picture according to the Color Key.

Color Key
7: Blue   **6**: Brown   **5**: Green   **4**: Gray   **3**: Yellow

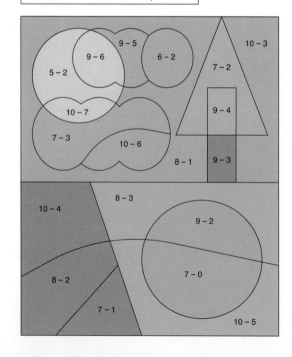

---

5 Catch the bugs in containers by drawing lines to match.

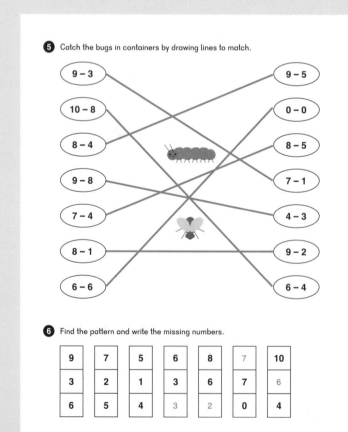

9 – 3   9 – 5

10 – 8   0 – 0

8 – 4   8 – 5

9 – 8   7 – 1

7 – 4   4 – 3

8 – 1   9 – 2

6 – 6   6 – 4

6 Find the pattern and write the missing numbers.

9	7	5	6	8	7	10
3	2	1	3	6	7	6
6	5	4	3	2	0	4

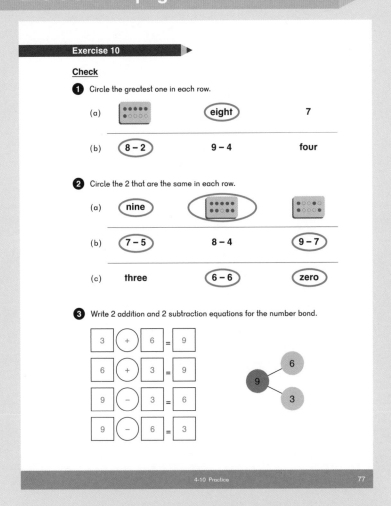

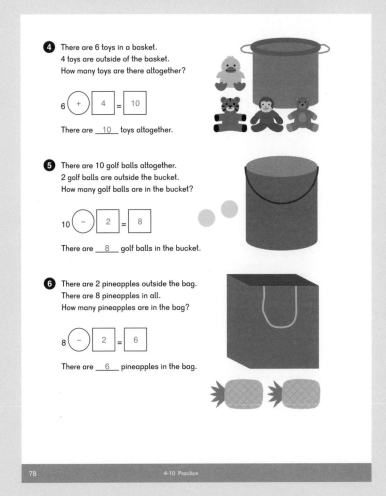

Exercise 11

Check

1 Circle the greatest one in each row.

(a) $(7 + 1)$ five   7 – 1

(b) 5 + 2   9 – 2 $(4 + 4)$

(c) 2 + 3 $(9 – 3)$ 8 – 3

(d) 9 + 0   8 – 2 $(5 + 5)$

2 What number is...

2 more	
6 + 2	10
5 – 4	3
7 – 0	9

2 less	
3 + 4	5
5 + 5	8
7 – 5	0

3 Write the missing numbers.

4	3	2	1	0

6	7	8	9	10

4 Write + or – in ◯ and a number in ☐ to make the equations true.

5 ⊕ 4 = 9   5 ⊖ 2 = 3

7 ⊖ 4 = 3   7 ⊕ 3 = 10

9 ⊖ 7 = 2   2 ⊕ 7 = 9

8 ⊖ 4 = 4   6 ⊕ 2 = 8

5 There are 4 lizards in the bush.
There are 3 lizards not in the bush.
How many lizards are there in all?

4 ⊕ 3 = 7

There are __7__ lizards in all.

6 Sofia wants to draw 9 hearts.
She has drawn 4 hearts.
How many more hearts does she need to draw?

9 ⊖ 4 = 5

She needs to draw __5__ more hearts.

7 Dion has 10 mangoes.
After he gives some away, he has 7 mangoes left.
How many mangoes did he give away?

10 ⊖ 7 = 3

He gave __3__ mangoes away.

8 There were 8 goats.
Some ran away.
2 goats are left.
How many goats ran away?

8 ⊖ 2 = 6

__6__ goats ran away.

Challenge

9 (a) 2 more than 10 – 3 is __9__.

(b) __4__ is between 8 – 5 and 7 – 2.

(c) __2__ less than 9 – 7 is 0.

10 (a) 9 – 2 = 3 + 4   (b) 2 + 4 = 10 – 4

(c) 8 – 4 = 1 + 3   (d) 1 + 6 = 10 – 3

11 Fill in the boxes using 1, 1, 2, 2, 3, 5, 6, 7, 9.
Subtract across and down.

9	–	7	=	2
–		–		–
6	–	5	=	1
=		=		=
3	–	2	=	1

Hint: Start with 9 in the top left box, and 1 in the lower right.

Or

9 – 6 = 3
7 – 5 = 2
2 – 1 = 1

Teacher's Guide 1A Chapter 4

Suggested number of class periods: 7 – 8

Lesson		Page	Resources		Objectives
	Chapter Opener	p. 121	TB:	p. 83	Investigate counting to 20.
1	Numbers to 20	p. 122	TB: WB:	p. 84 p. 83	Understand the structure of a two-digit number within 20 as 10 and some more.
2	Add or Subtract Tens or Ones	p. 124	TB: WB:	p. 86 p. 85	Relate the digits in a two-digit number within 20 to the part-whole meaning of a number bond. Write equations that recognize two-digit numbers as the sum of tens and ones.
3	Order Numbers to 20	p. 126	TB: WB:	p. 89 p. 87	Order numbers from 11 to 20 from least to greatest and greatest to least.
4	Compare Numbers to 20	p. 129	TB: WB:	p. 92 p. 89	Compare numbers to 20.
5	Addition	p. 131	TB: WB:	p. 94 p. 91	Add within 20 without regrouping by adding the ones.
6	Subtraction	p. 133	TB: WB:	p. 96 p. 95	Subtract within 20 without regrouping by subtracting from the ones.
7	Practice	p. 135	TB: WB:	p. 98 p. 99	Practice comparing, ordering, adding, and subtracting numbers to 20.
	Workbook Solutions	p. 137			

Chapter 5: Numbers to 20 reviews and extends counting numbers from 11 to 20 by emphasizing place value of "ten and some more." (**Dimensions Math® Kindergarten B Chapter 7: Numbers to 20** also introduces numbers to 20.) Building on **Chapter 1: Numbers to 10**, students will learn to:

- Count, read, and write whole numbers to 20.
- Compare numbers within 20 using comparison language: "greater/more than," "less/fewer than," or "the same as." (See **Chapter 1** for comparison vocabulary.)
- Order and compare numbers to 20.
- Add and subtract within 20 with no regrouping.

Students often struggle with 11, 12, and 13. They need to hear, speak, and write the teens to ensure they aren't confused with tens (13 vs. 31). Identifying teen numbers as 10 and some more will also clear up student confusion.

Lesson 4: Comparing Numbers to 20 uses language in place of the symbols < and >, which will be introduced in **Dimensions Math® 2A**. Both parts of a comparison should be identified:

7 is more than 5.

I have more cats than dogs.

Instead of:

7 is more.

I have more cats.

The latter examples are ambiguous. "I have more cats" could mean, "I have more cats than yesterday." One way to compare is to identify the amounts that are the same and then count on to see which number is greater or less. (See **Notes** from Chapter 1 for more on comparison language.) By the end of this chapter, students should be able to compare numbers directly without needing to count objects.

Lesson 5: Addition and **Lesson 6: Subtraction** introduce adding and subtracting within the ones place: 14 + 2, 17 − 5.

Materials

- Sidewalk chalk
- Paper plates
- Painter's tape
- Pebble or hopscotch marker
- Straws
- Linking cubes
- Two-color counters
- Plastic baggies
- Classroom items such as rubber bands, erasers, markers, or other manipulatives
- Playing cards
- Whiteboards

Note: Materials for Activities will be listed in detail in each lesson.

Blackline Masters

- Number Cards
- Number Word Cards
- Ten-frame Cards
- Blank Double Ten-frames
- Number Cards — Large
- Blank Ten-frame
- Teens Target — Addition Game Board
- Teens Target — Subtraction Game Board

Storybooks

- *Piglets Playing: Counting from 11 to 20* by Megan Atwood
- *Twenty Big Trucks in the Middle of the Street* by Mark Lee
- *Math for All Seasons* by Greg Tang
- *1 to 20, Animals Aplenty* by Katie Viggers

Letters Home

- Chapter 5 Letter

Notes

Chapter Opener

Objective

- Investigate counting to 20.

Students are probably able to count from 0 to 20 before this lesson. In the **Chapter Opener**, they can practice counting with objects in the classroom or with the activities included.

Teachers should assess each student's ability to count. **Lesson 1: Numbers to 20** will build on counting to 20 by looking at teen numbers as "10 and some more."

Have students look at the picture and discuss how many of each item can be seen. Ask which groups are easier to count and why.

Activity

▲ Hopscotch to 20

Materials: Pebble or other marker for each player, sidewalk chalk or paper plates and painter's tape

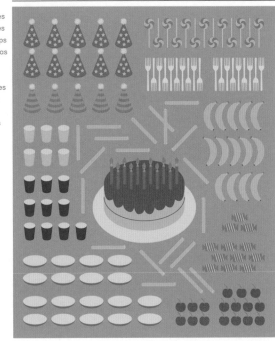

Chapter 5

Numbers to 20

11 candles	
12 candies	
13 lollipops	
14 bananas	
15 hats	
16 cups	
17 cherries	
18 plates	
19 forks	
20 straws	

83

83

20	
18	19
17	
15	16
14	
12	13
11	
10	

Home

Play outside or in a gym. Draw a hopscotch board using chalk, or tape down paper plates to create a hopscotch board with 10 as the starting spot.

Players take turns standing in the 10 square (Home) and tossing the marker. On their first turn, players aim for the 11 square. On each turn, players hop over the square with the marker and continue hopping in order, saying the numbers in each square aloud.

Square 20 is a rest stop. Players can put both feet down before turning around and hopping back to 10. Players pause in square 12 to pick up the marker from square 11, hop in square 11, and then out.

On her next turn, the player aims her marker for square 12, etc.

A player's turn is over if:

- Her marker does not land in the correct square.
- She loses her balance and puts a second foot down.
- She lands on a square where a marker is.

The winner is the first player to get through all 10 turns.

Lesson 1 Numbers to 20

Objective

- Understand the structure of a two-digit number within 20 as 10 and some more.

Lesson Materials

- Straws or linking cubes, 20 per student
- Number Cards (BLM) 11 to 20 for each student
- Ten-frame Cards (BLM) 11 to 20
- Number Word Cards (BLM) 11 to 20, 1 set per student

Think

Provide pairs of students between 11 and 20 straws or linking cubes and have them count them. Model counting 1, 2, 3, 4, etc.

Ask students if there is an easier way to count the items and allow them to share suggestions. Examples:

- Count by 2s.
- Count by 5s.
- Group them up into bundles of 10.

Show students that it can be easy to group 10 and count from there. Have students make a 10 (either by linking their cubes or bundling the straws) and count on from the 10.

In their sets of Number Cards (BLM) and Number Word Cards (BLM), have students find the number card and number word card that matches how many items they have.

Encourage students to think of their number as 10 and ___.

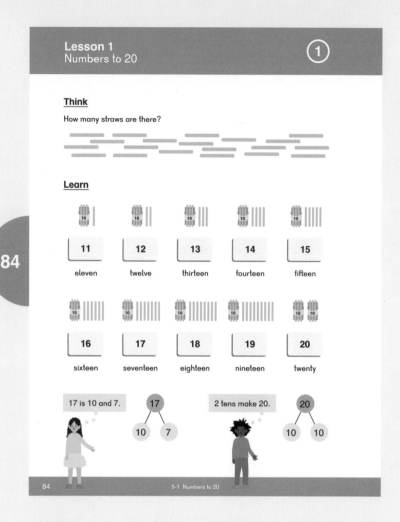

Learn

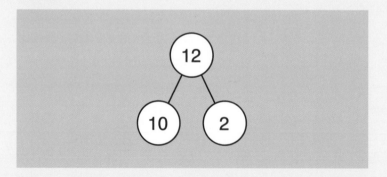

Discuss Sofia and Dion's thoughts and number bonds.

Write additional number bonds on the board similar to those at the bottom of the page, with 10 as one of the parts.

- 11 is 10 and 1.
- 12 is 10 and 2.

Do

1 Have students match Number Cards (BLM), Number Word Cards (BLM), and Ten-frame Cards (BLM) for numbers 11 to 20.

Activities

▲ Magic Thumb

Using your thumb to point up or down, have students chorally count on and back within 20 by ones.

Example: "Let's count by ones starting at 10, first number?" Class: "10." Point thumb up (class responds, "11"), then point up again (class responds, "12"). Point down (class responds, "11"), and so on.

▲ Ten and More Face-off

Materials: 4 sets of Ten-frame Cards (BLM) 0 to 9, Ten-frame Card (BLM) for 10 for each player

Play in pairs or groups of three or four.

Each player receives a Ten-frame Card (BLM) for 10 and places it faceup in front of herself. This "10" card becomes one of the addends in each face-off.

Deal out the remaining Ten-frame Cards (BLM) and have each student place their pile facedown in front of them. Each player turns over the top card from their pile and adds that card to the 10, saying the addition problem (for example, "10 plus 8 equals 18").

The player with the greatest total wins all of the non-10 cards. All players retain their original "10" card for their next face-off.

The player with the most cards at the end wins.

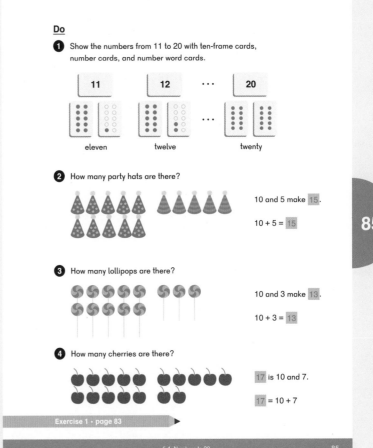

Exercise 1 • page 83

Lesson 2 Add or Subtract Tens or Ones

Objectives

- Relate the digits in a two-digit number within 20 to the part-whole meaning of a number bond.
- Write equations that recognize two-digit numbers as the sum of tens and ones.

Lesson Materials

- Blank Double Ten-frames (BLM), 1 per student
- Two-color counters, 20 per student

Think

Using the textbook or objects in the classroom, pose the problem with the cups.

"How many [objects] in all?"

Discuss student solutions.

Learn

Give students Blank Double Ten-frames (BLM) and counters. Have them fill one of the ten-frames with one color of the counters. Ask, "How may counters are there?" Students should respond, "10."

Ask, "How do you know?" Students may answer:

- It has 10 squares and they all have a counter, so it's showing a 10.
- I counted them.

Have students put 6 of the other color counters on their empty ten-frame. Ask, "How many counters are on both ten-frames?" Possible answers are:

- 16
- 10 and 6 make 16.

Show students an empty number bond and fill in the parts. Below the number bond, add the equation "10 + 6 = ?".

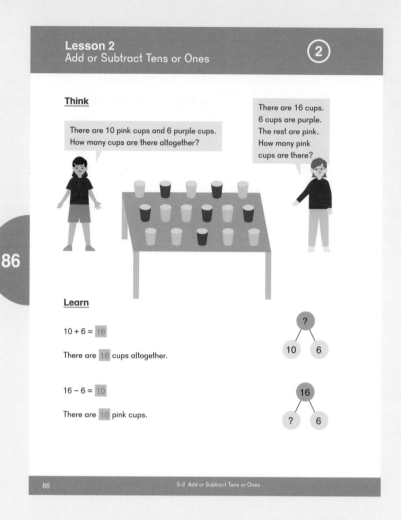

Have students make the number 17 on their ten-frames. Ask them, "If we have 17 counters and we take away 7, where would we write these numbers on the number bond?" A student may say, "Put the 17 in the whole, and put 7 in a part."

Ask students to show a subtraction equation for 17 minus 7.

Finally, ask, "What is 17 − 7?"

Have students discuss **Learn** in the textbook. How are the problems similar to what they did with their ten-frames? How are they different?

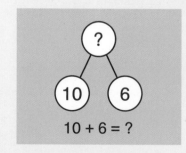

10 + 6 = ?

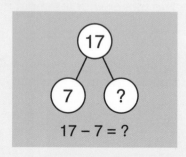

17 − 7 = ?

Do

Students should note that the order of the parts in addition does not matter, however the order in subtraction does.

5 (f) & (h) Some students may struggle to understand why they are adding in a subtraction problem. Have them recall part-whole and that they need to find the whole from the two parts.

To extend, provide students with more abstract word problems of varying difficulty. Have them write the equation and then solve the problem. For example:

- I think of a number, then I subtract 7. The result is 10. What is my original number?
- What number, when added to 8, makes 18?
- What number, when it is subtracted from 17, makes 10?
- Together Bob and Ann have 16 crayons. If Bob has 4 more crayons than Ann, how many crayons does Bob have?

Exercise 2 • page 85

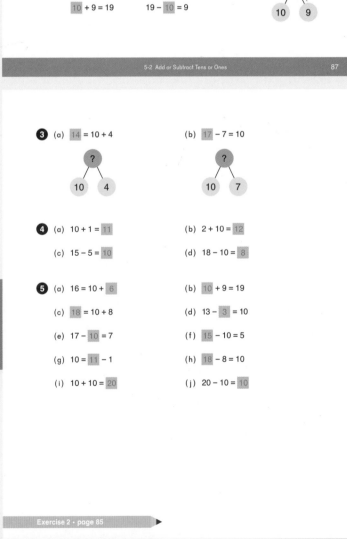

Lesson 3 Order Numbers to 20

Objective

- Order numbers from 11 to 20 from least to greatest and greatest to least.

Lesson Materials

- 200 linking cubes
- Number Cards (BLM) from 0 to 20 — 1 card per student

Think

Provide students (or pairs of students) with some linking cubes and a Number Card (BLM). Have them create a linking cube tower with the number of linking cubes shown on their card.

Have students lay out their cards and towers on the floor or table in order from least to greatest. This will use 200 cubes. Students can be called in order beginning with 0 to put their towers in order.

To extend, call a random number and have students put their towers on a line where they think they belong in the sequence.

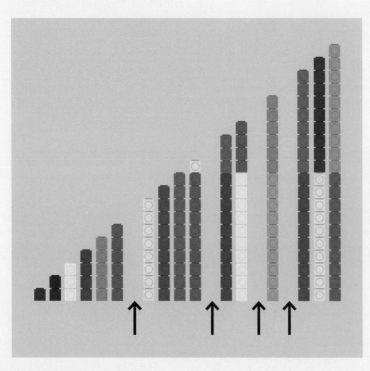

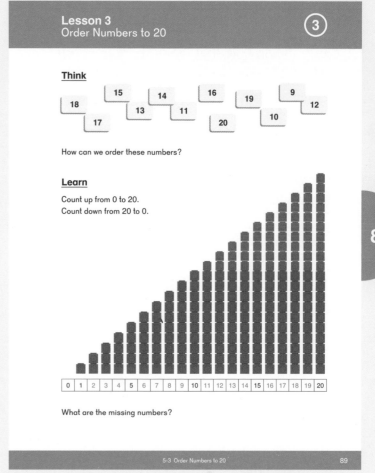

Lesson 3
Order Numbers to 20 ③

Think

18   15   14   16   19   9
17   13   11   20   10   12

How can we order these numbers?

Learn

Count up from 0 to 20.
Count down from 20 to 0.

| 0 | 1 | 2 | 3 | 4 | 5 | 6 | 7 | 8 | 9 | 10 | 11 | 12 | 13 | 14 | 15 | 16 | 17 | 18 | 19 | 20 |

What are the missing numbers?

Learn

Have students compare the towers they ordered with the picture in the textbook.

Ask students, "By how many cubes are the towers increasing or decreasing each time we add a number?" and, "What do you notice about the blocks in the book?"

Potential student answers:

- The numbers go up (down) by 1.
- The blocks are different colors after the first 10.
- The blocks are 10 and some more.

Do

❶ — **❷** Students will use their Number Cards (BLM) to complete these problems.

❸ Students should be able to find the answer by looking at the textbook. Students who struggle should continue to use their Number Cards (BLM).

❺ — **❼** Extend these problems by having students work in pairs and play the activities below.

❽ The **Greater Than/Less Than** activity on the following page is similar to prior lessons. Students now play with number cards, which are more abstract than ten-frame cards.

Activities

▲ Put Yourself on the Line

Materials: Number Cards — Large (BLM) 0 to 20

Pass out Number Cards — Large (BLM) to students. Have them put themselves in order in front of the class without talking.

★ To play in teams, put students in groups of four or five and give each of them a Number Card — Large (BLM) at random. Each team puts themselves in order from least to greatest or vice versa.

The winning team is the one that gets themselves in order first.

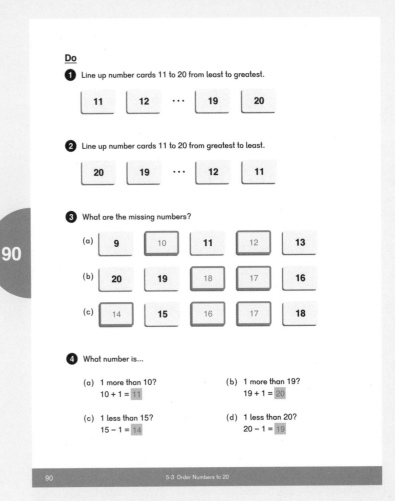

90

▲ Greater Than/Less Than

Materials: Number Cards (BLM) 11 to 20

Player 1 shows a Number Card (BLM) 11 to 20. Player 2 has to find the number card or say the numbers that are 2 more and 2 less than Player 1's card.

★ Play 3 more or 3 less, or 10 more or 10 less.

▲ What are the Missing Numbers?

Materials: Number Cards (BLM) 11 to 20

Player 1 has the set of Number Cards (BLM) 11 to 20 and removes 2 random cards. Player 2 takes the cards and figures out which numbers are missing.

Solitaire option: Shuffle numbers 0 to 20, or 11 to 20. Student turns over the top 5 and puts them in order from least to greatest. Draw remaining numbers 1 at a time and insert them in the correct places.

Exercise 3 • page 87

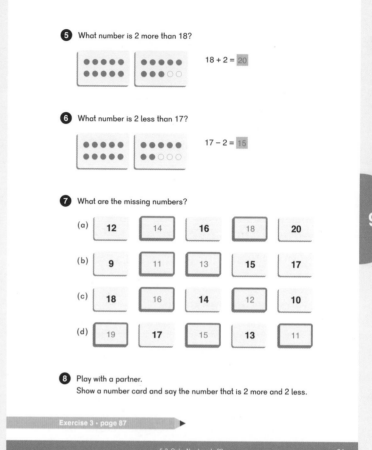

5 What number is 2 more than 18?

$18 + 2 = 20$

6 What number is 2 less than 17?

$17 - 2 = 15$

7 What are the missing numbers?

(a) 12 14 16 18 20

(b) 9 11 13 15 17

(c) 18 16 14 12 10

(d) 19 17 15 13 11

8 Play with a partner.
Show a number card and say the number that is 2 more and 2 less.

Exercise 3 • page 87

Lesson 4 Compare Numbers to 20

Objective

- Compare numbers to 20.

Lesson Materials

- Bags containing between 11 and 20 each of two kinds of manipulatives (rubber bands, cubes, bears, erasers, markers, counters, etc.), 1 per pair of students
- Blank Double Ten-frames (BLM)

Think

Provide each pair of students with a bag of manipulatives. Each student chooses one type of the items in the bag and counts how many they have using a Blank Double Ten-frame (BLM).

Have partners compare who has the most and who has the least. Have each pair of students share how many of each item they have. Ask how they know who has more (or less).

Have students share their bags by saying:

- There are more ____ than ____.
- There are fewer ____ than ____.

Ensure that students include both objects.

For example, if one bag has 16 paper clips and 13 pom-poms, students would say:

- There are more paper clips than pom-poms. There are fewer pom-poms than paper clips.
- 6 is greater than 3 so 16 is greater than 13. 3 is fewer than 6 so 13 is fewer than 16.

Learn

Have students discuss the example with the bananas and candies. Ask them to finish Alex's thought, "... so 14 is greater than 12."

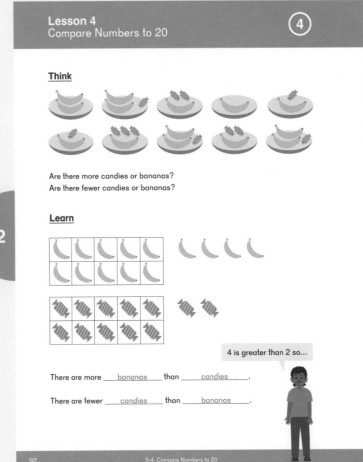

Do

Extend problems by asking students, "How many more or how many fewer?"

Activity

▲ More Face-off or Fewer Face-off

Materials: Number Cards (BLM) 1 to 20

Players each flip over a Number Card (BLM) at the same time. The greatest or least number card wins (depending on version of game).

Students can use linking cubes or counters to see whose number is greater.

★ Have the winner say how many greater or less their card is than the other player's card.

Exercise 4 • page 89

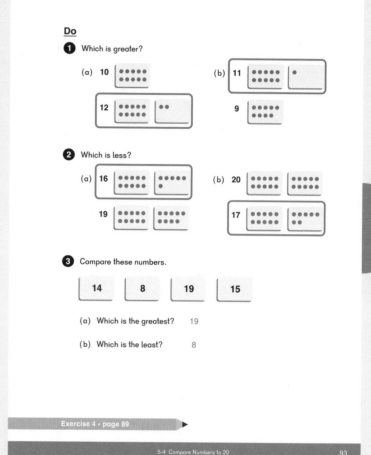

Lesson 5 Addition

Objective

- Add within 20 without regrouping by adding the ones.

Lesson Materials

- Two-color counters, 20 per student
- Blank Ten-frames (BLM)

Think

Pose the problem with the candies.

Provide students with two-color counters and a Blank Ten-frame (BLM). Have students share their strategies for solving the problem.

Possible solutions:

- I counted all of the candies.
- I put 10 on the ten-frame and counted 5 more.
- I started with 12 and counted on 3.

Note: Students are not expected to use formal place value language at this time.

Learn

Ask students which strategy Dion used to solve the problem. Emphasize that Dion added the loose candies and did not change the 10 on the ten-frame. Ask them how they could use this strategy for other problems. Have them practice with problems of teen numbers and ones:

14 + 5

17 + 2

11 + 6, etc.

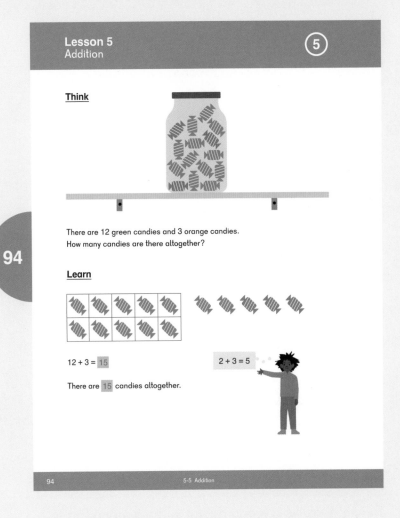

Lesson 5
Addition ⑤

Think

There are 12 green candies and 3 orange candies. How many candies are there altogether?

Learn

12 + 3 = 15

There are 15 candies altogether.

2 + 3 = 5

94   5-5 Addition

94

Remind students that we can add ones to a teen number by splitting the teen number into tens and ones, then adding together the ones. Have students think about how this would look with a number bond. For the candies:

- Split the whole 12 into parts 10 and 2.
- Add parts 3 and 2 to make a whole of 5.
- Add parts 10 and 5 to make the whole of 15.

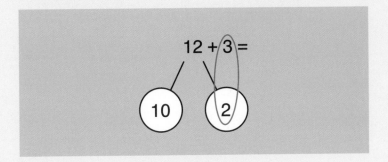

Do

Have students complete these problems on a whiteboard and share their strategies.

❷ (b) Remind students that they should use a number bond that facilitates their addition problem. 9 + 10 is just as correct as 10 + 9.

Activity

▲ Teens Target — Addition

Materials: Teens Target — Addition Game Board (BLM) for each player, and playing cards, Number Cards (BLM), or Ten-frame Cards (BLM) 0 to 9

Players take turns drawing cards and writing numbers in empty squares on th Teens Target — Addition Game Board (BLM). If they cannot play, they discard that card.

Example game play:

Player 1 draws a 6 and puts it in the second row in either empty square. On her next turn, she draws a 4. The 4 could go in the first equation, but not in the second equation as that will make the equation false.

The first player to make five true equations wins.

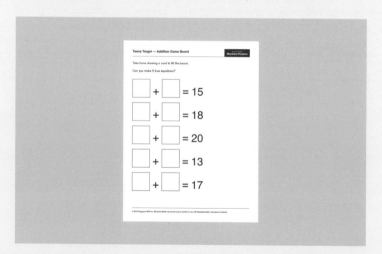

Exercise 5 • page 91

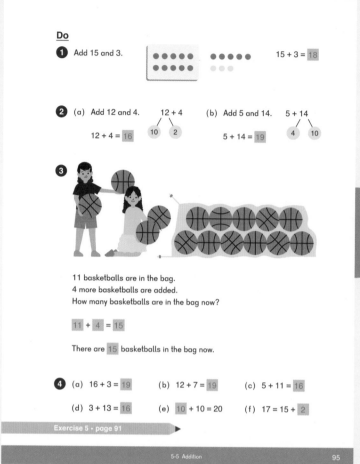

Do

❶ Add 15 and 3. $15 + 3 = \boxed{18}$

❷ (a) Add 12 and 4. $12 + 4$
   $12 + 4 = \boxed{16}$ ⑩ ②

(b) Add 5 and 14. $5 + 14$
   $5 + 14 = \boxed{19}$ ④ ⑩

❸

11 basketballs are in the bag.
4 more basketballs are added.
How many basketballs are in the bag now?

$\boxed{11} + \boxed{4} = \boxed{15}$

There are $\boxed{15}$ basketballs in the bag now.

❹ (a) $16 + 3 = \boxed{19}$ (b) $12 + 7 = \boxed{19}$ (c) $5 + 11 = \boxed{16}$

(d) $3 + 13 = \boxed{16}$ (e) $\boxed{10} + 10 = 20$ (f) $17 = 15 + \boxed{2}$

Exercise 5 • page 91

5-5 Addition   95

Lesson 6 Subtraction

Objective

- Subtract within 20 without regrouping by subtracting from the ones.

Lesson Materials

- Two-color counters, 20 per student
- Blank Ten-frames (BLM)

Think

Pose the problem with the foxes.

Provide students with two-color counters and a Blank Ten-frame (BLM). Have them share their strategies for solving the problem.

Possible solutions:

- I took 3 away.
- I put 10 on the ten-frame and had 5 loose, and took away 3 from the 5.
- I started with 15 and counted back 3.

Note: Formal place value language is not used at this point.

Learn

Ask students which strategy Sofia used to solve the problem. Remind students that we can subtract ones from a teen number by decomposing the teen number into tens and ones and then subtracting the ones from the ones.

Split 15 into 10 and 5.
Subtract 3 ones from 5 ones to make 2 ones.
Add 10 and 2 ones to make 12.

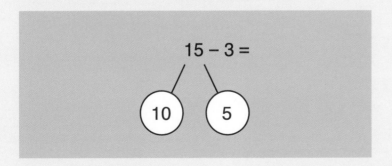

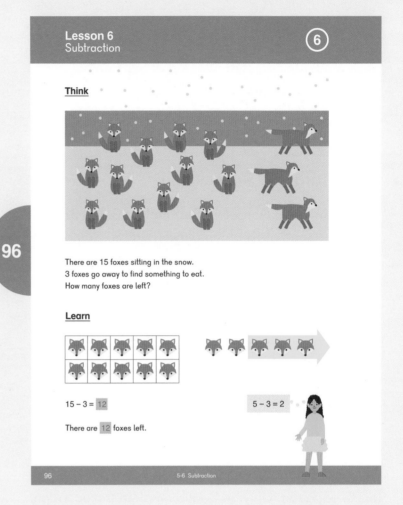

Lesson 6
Subtraction
6

Think

There are 15 foxes sitting in the snow.
3 foxes go away to find something to eat.
How many foxes are left?

Learn

15 − 3 = 12

There are 12 foxes left.

5 − 3 = 2

96   5-6 Subtraction

Have students practice with problems of teen numbers and ones:

$19 - 5$
$12 - 2$
$16 - 0$, etc.

Have students show the different number bonds.

Do

Have students complete these problems on a whiteboard, then discuss their strategies.

Activity

▲ Teens Target — Subtraction

Materials: Teens Target — Subtraction Game Board (BLM) for each player, and playing cards, Number Cards (BLM), or Ten-frame Cards (BLM) 0 to 9

Players take turns drawing cards and writing the numbers in empty squares on the Teens Target — Subtraction Game Board (BLM). If they cannot play, they discard that card.

Example game play:

Player 1 draws a 3 and puts it in the first row to make $13 - \square = 10$. Only a 3 will complete that equation. On her next turn, she draws a 5 and puts it in the fourth equation. Only an 8 will complete the equation. The first player to make five true equations wins.

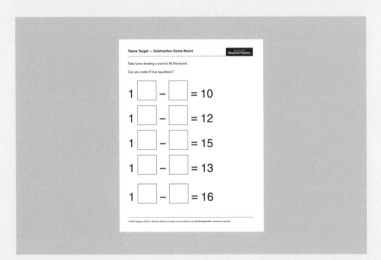

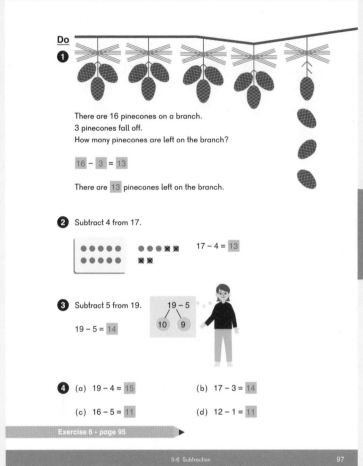

Do

1 There are 16 pinecones on a branch.
3 pinecones fall off.
How many pinecones are left on the branch?

$16 - 3 = 13$

There are 13 pinecones left on the branch.

2 Subtract 4 from 17.

$17 - 4 = 13$

3 Subtract 5 from 19.

$19 - 5 = 14$

$19 - 5$
$10 \quad 9$

4 (a) $19 - 4 = 15$ (b) $17 - 3 = 14$

    (c) $16 - 5 = 11$ (d) $12 - 1 = 11$

Exercise 6 • page 95

5-6 Subtraction   97

Exercise 6 • page 95

Lesson 7 Practice

Objective

- Practice comparing, ordering, adding, and subtracting numbers to 20.

Lesson Materials

- Counters

Practice

After students complete the **Practice** in the textbook, have them continue working with numbers to 20 by playing games from this chapter.

Chapter 6: Addition to 20 and **Chapter 7: Subtraction Within 20** will introduce methods of adding and subtracting within 20 with regrouping. This includes adding 2 one-digit numbers when the sum is greater than 10, or subtracting a one-digit number from a number between 10 and 20 when the difference is less than 10.

Activities

▲ **Target 20**

Materials: Playing cards, Number Cards (BLM), or Ten-frame Cards (BLM) 0 to 10

This game can be played in a small group or individuals can challenge themselves. The goal is to get as close to 20 as possible without going over.

Player 1 holds the deck of cards and turns over one card at a time, adding as he turns over the cards. Player 1 can stop any time or choose to add another card. When Player 1 stops at a sum they believe will be closest to 20, he passes the remaining deck to the next player. Players who go over 20 are out of the round. When all players have had a turn, the round is over. The player closest to 20 without going over wins.

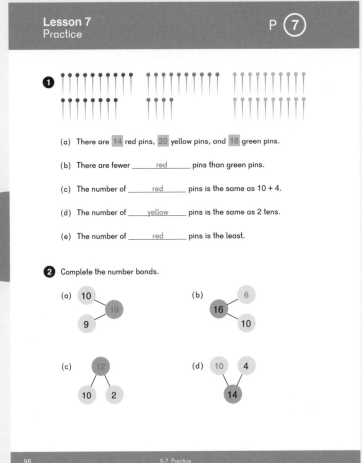

★ 20 or Bust

Materials: Playing cards, Number Cards (BLM), or Ten-frame Cards (BLM) 0 to 10

Game plays like blackjack. Face cards are worth 10 and aces can be either 1 or 11.

Players take turns being the Dealer. The Dealer gives two cards to each player and then asks each player in turn if they wish to receive another card (hit) or not receive another card because they don't want to go over 20 (stand).

If a player goes over 20, he "busts" and is out that round.

The winner keeps cards. The player with the most cards at the end wins.

Brain Works

★ Order Up!

Materials: Number Cards (BLM) 1 to 20

Provide students with Number Cards (BLM) listed below. Have them find the two missing cards and put the cards in order.

- 9, 13, 15, 19 (missing 11 and 17)
- 20, 18, 14, 10 (missing 16 and 12)
- 4, 7, 10, 19 (missing 13 and 16)
- 19, 18, 17, 15 (missing either 16 and 14, or 20 and 16)

Exercise 7 • page 99

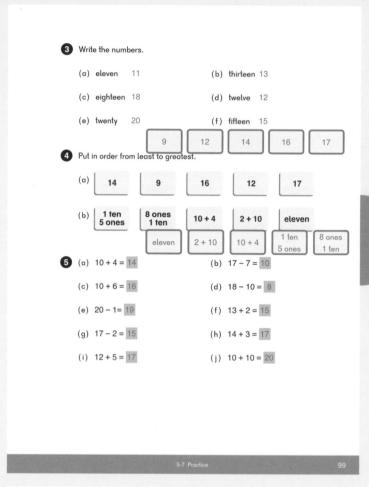

③ Write the numbers.

(a) eleven   11   (b) thirteen   13

(c) eighteen   18   (d) twelve   12

(e) twenty   20   (f) fifteen   15

④ Put in order from least to greatest.

(a) 14   9   16   12   17

   9   12   14   16   17

(b) 1 ten / 5 ones   8 ones / 1 ten   10 + 4   2 + 10   eleven

   eleven   2 + 10   10 + 4   1 ten / 5 ones   8 ones / 1 ten

⑤ (a) 10 + 4 = 14   (b) 17 − 7 = 10

(c) 10 + 6 = 16   (d) 18 − 10 = 8

(e) 20 − 1 = 19   (f) 13 + 2 = 15

(g) 17 − 2 = 15   (h) 14 + 3 = 17

(i) 12 + 5 = 17   (j) 10 + 10 = 20

5-7 Practice   99

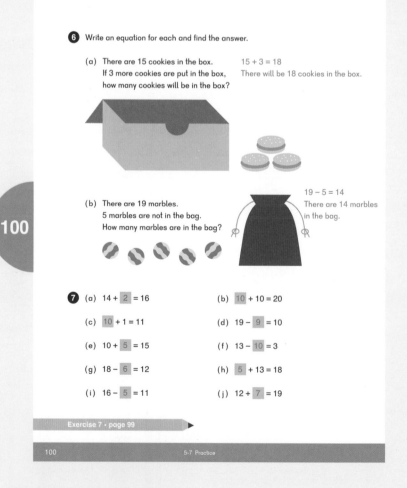

⑥ Write an equation for each and find the answer.

(a) There are 15 cookies in the box.   15 + 3 = 18
If 3 more cookies are put in the box,   There will be 18 cookies in the box.
how many cookies will be in the box?

(b) There are 19 marbles.   19 − 5 = 14
5 marbles are not in the bag.   There are 14 marbles
How many marbles are in the bag?   in the bag.

⑦ (a) 14 + 2 = 16   (b) 10 + 10 = 20

(c) 10 + 1 = 11   (d) 19 − 9 = 10

(e) 10 + 5 = 15   (f) 13 − 10 = 3

(g) 18 − 6 = 12   (h) 5 + 13 = 18

(i) 16 − 5 = 11   (j) 12 + 7 = 19

Exercise 7 • page 99

100   5-7 Practice

Chapter 5 Numbers to 20

Exercise 1

Basics

1 Circle groups of 10.
Then write the missing numbers.

Which 10 a student circles may vary.

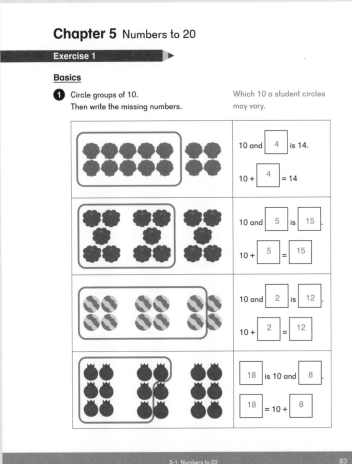

10 and 4 is 14.

10 + 4 = 14

10 and 5 is 15

10 + 5 = 15

10 and 2 is 12

10 + 2 = 12

18 is 10 and 8.

18 = 10 + 8

Practice

2 (a) 19 — 10, 9

(b) 13 — 10, 3

(c) 11 — 10, 1

3 Circle 10.
Write how many in all.

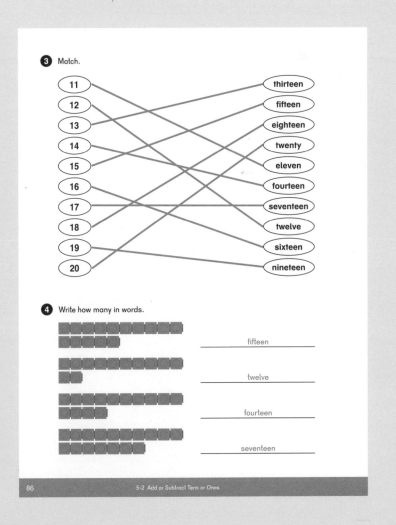

14

16

18

19

11

20

15

13

Exercise 2

Basics

1

10 + 6 = 16   6 + 10 = 16

16 – 6 = 10   16 – 10 = 6

16 — 10, 6

Practice

2 (a) 10 + 8 = 18   (b) 11 – 1 = 10

(c) 9 + 10 = 19   (d) 17 – 10 = 7

(e) 10 + 4 = 14   (f) 10 + 3 = 13

(g) 15 – 5 = 10   (h) 12 – 10 = 2

3 Match.

11 — fourteen
12 — seventeen
13 — fifteen
14 — eleven
15 — eighteen
16 — sixteen
17 — twelve
18 — twenty
19 — thirteen
20 — nineteen

4 Write how many in words.

fifteen

twelve

fourteen

seventeen

Exercise 3

Basics

❶

1	2	3	4	5	6	7	8	9	10
11	12	13	14	15	16	17	18	19	20

(a) 1 more than 10 is __11__.

(b) __16__ is 1 more than 15.

(c) 19 is 2 more than __17__.

(d) __12__ is 2 less than 14.

(e) 9 is 3 less than __12__.

(f) 3 more than __17__ is 20.

Practice

❷ Complete the number patterns.

(a) **10** | 12 | **14** | 16 | 18 | **20**

(b) 17 | **15** | 13 | **11** | **9** | 7

(c) **3** | **6** | 9 | **12** | 15 | 18

5-3 Order Numbers to 20   87

---

88   5-3 Order Numbers to 20

❸ Find and circle the number words in the puzzle.
The words go across or down.

- ten   • sixteen
- eleven   • seventeen
- twelve   • eighteen
- thirteen • nineteen
- fourteen • twenty
- fifteen

n	t	e	f	f	t	e	n	s	s
n	h	e	o	i	w	i	i	i	e
e	i	r	u	f	t	g	n	x	l
v	r	e	r	t	w	h	e	t	e
i	t	l	t	e	e	t	t	e	v
n	e	e	e	n	e	e	e	e	e
s	e	v	e	n	t	e	e	n	x
e	n	e	n	h	y	n	n	r	t
e	e	n	t	w	e	l	v	e	v
t	n	n	i	s	h	f	e	n	e

Exercise 4

Basics

❶ Write how many are in each set.

A 18   **B** 15

Set __A__ has more tomatoes than Set __B__.

18 is greater than 15.

C 13   **D** 12

Set __D__ has fewer thimbles than Set __C__.

12 is less than 13.

❷ Circle the greatest number.

(a) 12 **⑱** 15

(b) 11 9 **⑰** 13

❸ Circle the least number.

(a) 16 **⑬** 19

(b) 20 14 17 **⑫**

5-4 Compare Numbers to 20   89

---

90   5-4 Compare Numbers to 20

Practice

❹ Write the numbers in the box in order, from least to greatest.

19 12 20 17 8 | 8 | 12 | 17 | 19 | 20 |

❺ Write a number from the box to make each sentence true.

13   11
18   16   15

(a) 18 is the greatest number.

(b) 11 is the least number.

(c) 13 is greater than 11 and less than 15.

(d) 16 is less than 18 and greater than 15.

Challenge

❻ Circle the numbers in the box that are less than 17 and greater than 12.

20 **⑬** **⑮** 17 **⑯** 18 12

❼ Write the numbers from greatest to least.

| 1 ten 4 ones | eleven | 2 more than 13 | 1 less than 18 |

| 17 | 15 | 14 | 11 |

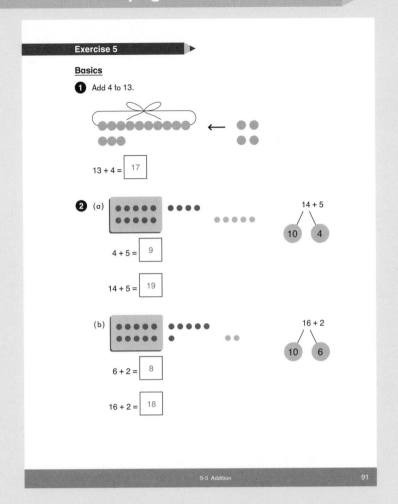

Exercise 5

Basics

1 Add 4 to 13.

$13 + 4 = \boxed{17}$

2 (a)

$14 + 5$

$10 \quad 4$

$4 + 5 = \boxed{9}$

$14 + 5 = \boxed{19}$

(b)

$16 + 2$

$10 \quad 6$

$6 + 2 = \boxed{8}$

$16 + 2 = \boxed{18}$

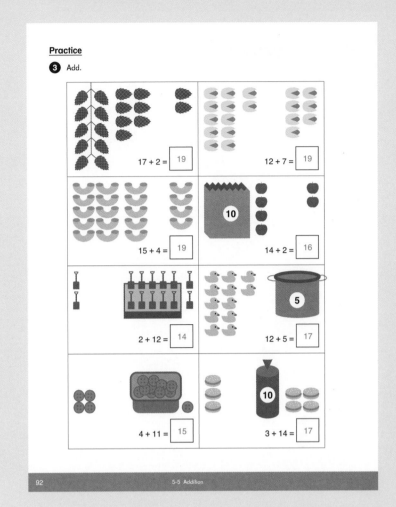

Practice

3 Add.

$17 + 2 = \boxed{19}$ $12 + 7 = \boxed{19}$

$15 + 4 = \boxed{19}$ $14 + 2 = \boxed{16}$

$2 + 12 = \boxed{14}$ $12 + 5 = \boxed{17}$

$4 + 11 = \boxed{15}$ $3 + 14 = \boxed{17}$

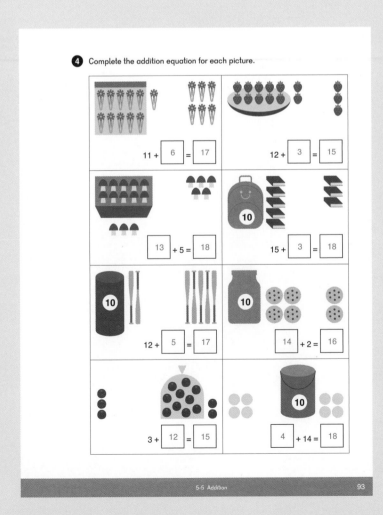

4 Complete the addition equation for each picture.

$11 + \boxed{6} = \boxed{17}$ $12 + \boxed{3} = \boxed{15}$

$\boxed{13} + 5 = \boxed{18}$ $15 + \boxed{3} = \boxed{18}$

$12 + \boxed{5} = \boxed{17}$ $\boxed{14} + 2 = \boxed{16}$

$3 + \boxed{12} = \boxed{15}$ $4 + 14 = \boxed{18}$

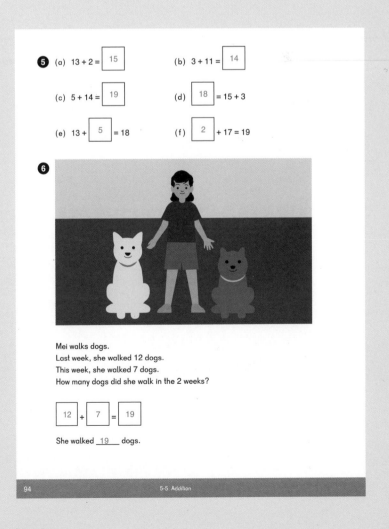

5 (a) $13 + 2 = \boxed{15}$ (b) $3 + 11 = \boxed{14}$

(c) $5 + 14 = \boxed{19}$ (d) $\boxed{18} = 15 + 3$

(e) $13 + \boxed{5} = 18$ (f) $\boxed{2} + 17 = 19$

6

Mei walks dogs.
Last week, she walked 12 dogs.
This week, she walked 7 dogs.
How many dogs did she walk in the 2 weeks?

$\boxed{12} + \boxed{7} = \boxed{19}$

She walked __19__ dogs.

   Teacher's Guide 1A Chapter 5

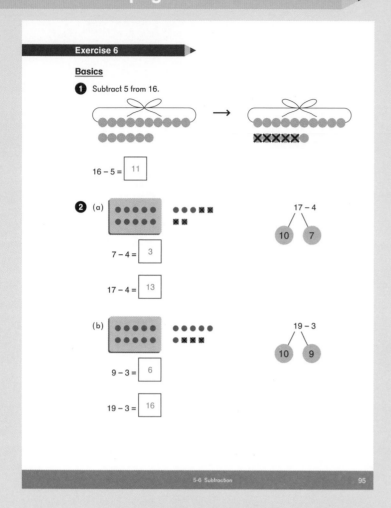

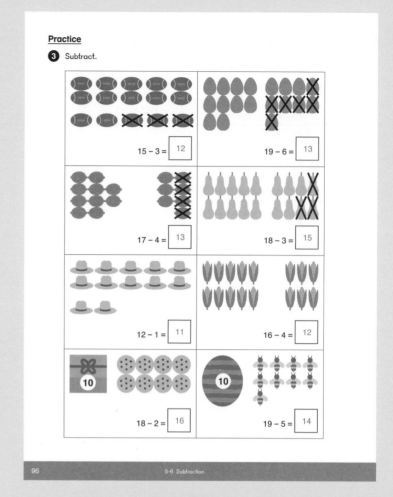

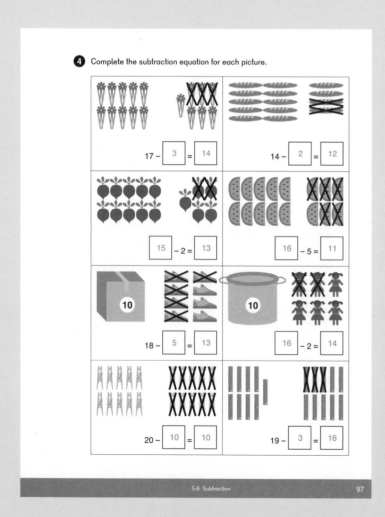

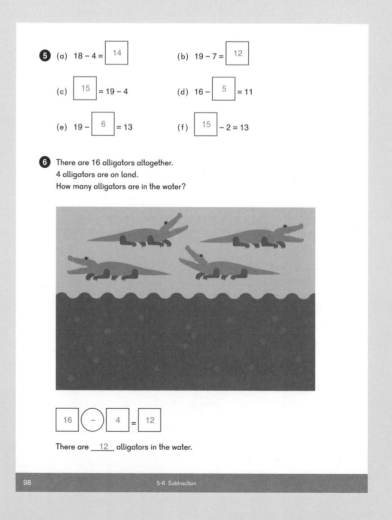

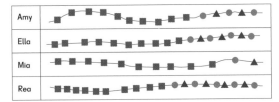

Check

1 Amy, Ella, Mia, and Rea each have some beads.

Amy	
Ella	
Mia	
Rea	

(a) _____Mia_____ has the fewest beads.

(b) _____Rea_____ has the most beads.

(c) _____Mia_____ has fewer beads than Amy.

(d) _____Ella_____ and _____Rea_____ have more beads than Amy.

(e) Write the number of beads each person has from least to greatest.

| 12 | 15 | 17 | 19 |

(f) Mia gets 4 more beads.

She now has __16__ beads.

(g) Rea loses 6 beads.

She now has __13__ beads.

2 Write the number that is...

1 more	
eleven	12
nineteen	20

1 less	
fifteen	14
twelve	11

3 (a) 10 + 7 = [17]   (b) 16 − 6 = [10]

(c) 12 + 7 = [19]   (d) 16 − 5 = [11]

4 Write the numbers.

| one more than eleven | 12 |
| two less than twenty | 18 |

| two more than thirteen | 15 |
| three less than seventeen | 14 |

Challenge

5 Write a check ✓ if the statement is correct.

1 ten and 5 ones is less than 18 ones and more than 13 ones.	✓
4 added to 12 is greater than 2 subtracted from 18.	
18 − 5 is 2 more than 6 + 5.	✓
8 + 5 is 1 less than 15 − 3.	

Notes

Suggested number of class periods: 5 – 6

Lesson		Page	Resources		Objectives
	Chapter Opener	p. 147	TB:	p. 101	Investigate counting and adding within 20.
1	Add by Making 10 — Part 1	p. 148	TB: WB:	p. 102 p. 101	Make 10 with 2 one-digit numbers by making a 10 with the first addend.
2	Add by Making 10 — Part 2	p. 151	TB: WB:	p. 105 p. 105	Make 10 with 2 one-digit numbers by making a 10 using the second addend.
3	Add by Making 10 — Part 3	p. 154	TB: WB:	p. 108 p. 109	Add 2 one-digit numbers with sums within 20 by making a 10 with either addend.
4	Addition Facts to 20	p. 157	TB: WB:	p. 113 p. 111	Learn addition facts to 20.
5	Practice	p. 159	TB: WB:	p. 115 p. 113	Practice addition facts to 20 to mastery.
	Workbook Solutions	p. 161			

Chapter 6: **Addition to 20** introduces students to addition of 2 one-digit numbers whose sum is between 10 and 20. The main strategy taught in this chapter is making a 10. That is, students will use number bonds to decompose, or split, addends into easier combinations to sum. Students should have mastered the addition facts to 10 before this chapter. If they have not, facts can be practiced and reinforced during the teaching of this chapter by playing the games and activities from prior lessons on a regular basis.

In **Chapter 3**: **Addition**, students learned to add by counting on 1, 2, or 3. This strategy is not re-taught in this chapter. Some students may use counting on.

Making 10 is a powerful strategy that supports the concept of place value and the concept of equality. These foundational skills will support standard algorithms and algebraic thinking in future years. While students are learning the strategy, they should be encouraged to represent how they decomposed the addends with number bonds.

In **Chapter 5**: **Numbers to 20**, students learned to add the ones to a teen number by decomposing the tens and ones in a two-digit number, then adding together the ones.

$$13 + 5 = 10 + 8 = 18$$

(10) (3)

$$7 + 5 = 10 + 2 = 12 \qquad 7 + 5 = 10 + 2 = 12$$

(3) (2)   (2) (5)

After **Chapter 6**, **Lesson 4**: **Addition Facts to 20**, students should be working towards fluency with the facts to 20. They should not be expected to explain every answer. Ultimately, the facts become automatic ("I just know that 6 + 8 is 14."); however, a strategy can be used when facts are not yet automatic.

There are other strategies that students may encounter like "doubles" and "doubles + 1." These strategies are not taught in the **Dimensions Math®** series as they do not generalize to larger numbers. For example, 98 + 98 is not easily learned as a doubles fact but if we subtract two from one addend and add it to the other we get 100 + 96 = 196.

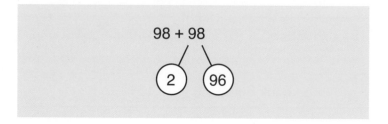

$$98 + 98$$

(2) (96)

Alternatively, students can add 100 + 100 and subtract 4.

Materials

- Dice
- Counters
- Linking cubes
- Playing cards
- Index cards
- 10-sided die, numbered 11 to 20
- Sidewalk chalk
- Painter's tape
- Whiteboards

Blackline Masters

- Blank Double Ten-frames
- Teens Target — Addition Game Board
- Number Cards
- Ten-frame Cards
- Spinner 11 – 20

Storybooks

- *The Mission of Addition* by Brian P. Cleary
- *Mission: Addition* by Loreen Leedy
- *12 Ways to Get to 11* by Eve Merriam
- *What's New at the Zoo?* by Suzanne Slade

Letters Home

- Chapter 6 Letter

Notes

Objective

- Investigate counting and adding within 20.

This **Chapter Opener** is an informal exploration of addition to 20. Have students look at the picture and discuss how many of each fruit or vegetable can be seen. Students may count them to find the answer to Sofia's question. Lead students to discuss other methods:

- Add the red and green tomatoes together to make 10, and then add the 5 yellow tomatoes to get 15 tomatoes in all.
- Add 5 watermelons and 5 watermelons to make 10, and then add 3 watermelons to get 13 watermelons in all.

Students who have these skills already will not need a full day of review.

This lesson may continue straight from **Think** to **Lesson 1**: **Add by Making 10 — Part 1**.

Activity

▲ Ten-frame Fill-up

Materials: 1 die per group or player, 1 Blank Double Ten-frame (BLM) per player, 20 counters per player

The game works best with two to four players. Players roll die to see who goes first. Player 1 rolls a die, and adds that many counters to one of the ten-frames on her Blank Double Ten-frame (BLM). Player 2 rolls and does the same. Play continues.

Players must fill a ten-frame with an exact roll. If adding counters to a ten-frame would make more than 10, the player passes his turn. The first player to fill both ten-frames wins.

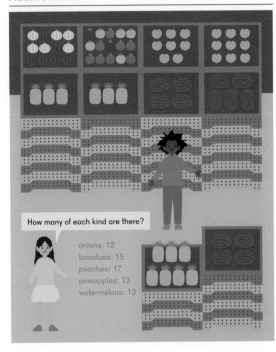

Chapter 6

Addition to 20

How many of each kind are there?

onions: 12
tomatoes: 15
peaches: 17
pineapples: 13
watermelons: 13

101

Objective

- Make 10 with 2 one-digit numbers by making a 10 with the first addend.

Lesson Materials

- Linking cubes, 10 each of 2 different colors per student
- Blank Double Ten-frames (BLM), 1 per student
- Peppers or classroom objects, 9 of one item and 4 of another

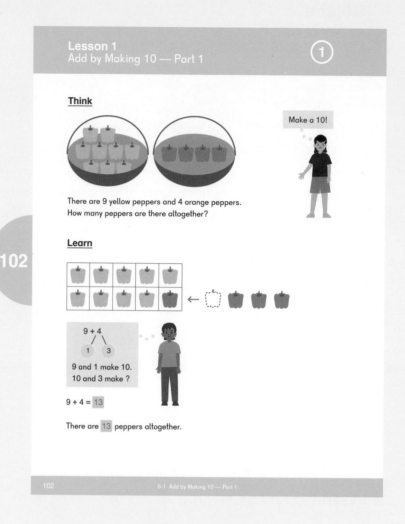

Think

Pose the pepper problem from **Think**. Provide students with linking cubes and a Blank Double Ten-frame (BLM) to help them work through the problem. Have students share the method they used for solving the problem. Examples:

- I counted them all.
- I counted on from 9.
- I used the ten-frames.

Learn

Have students begin by representing the yellow peppers on one ten-frame with one color of cubes, and the red peppers on the other ten-frame with the other color of cubes. Students should note that if they had one more pepper on their first ten-frame, that it would be a full ten-frame and easy to see that 10 and 3 make 13, the same as 9 and 4.

Help students realize that they are just moving the existing cubes between ten-frames to make a simpler problem, and that no cubes were added.

Provide additional examples by adding 9 and numbers less than 5. Follow with adding 8 and numbers less than 5.

Looking at the text, note how Alex is splitting the number 4 into 1 and 3.

Show the following number bond on the board, emphasize that the 9 and 1 are being combined to make a 10 by circling the numbers. Then adding the remaining part of 4, or 3.

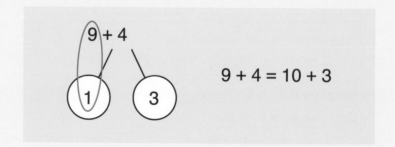

Do

While working through the **Do** problems, reinforce the concept that 8 plus 3 is the same as 10 plus 1 because we can take 2 from the 3 and add them to the 8. Now there are 10 and 1 more.

❶ — ❸ Students should work these problems on whiteboards. Have them show each part on a ten-frame as needed.

Students that struggle can cross off the number being split.

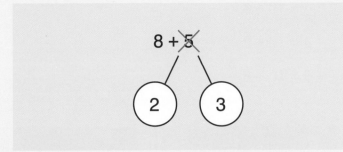

❹ Problems are scaffolded to encourage students to use the strategy. The first equation in a row encourages students to determine how many make 10 so they know how to decompose the second addend in the next equation. For example, 9 and 1 make 10, so 9 and 8 make 10 and ?

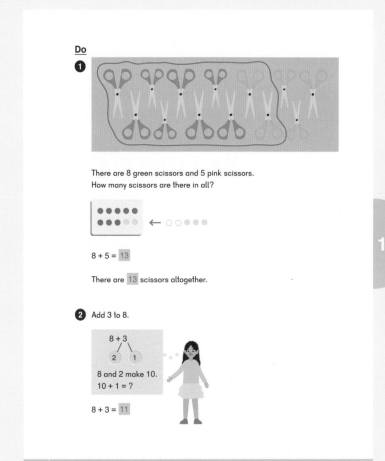

Do

❶

There are 8 green scissors and 5 pink scissors. How many scissors are there in all?

$8 + 5 =$ 13

There are 13 scissors altogether.

❷ Add 3 to 8.

8 + 3

2   1

8 and 2 make 10.
10 + 1 = ?

$8 + 3 =$ 11

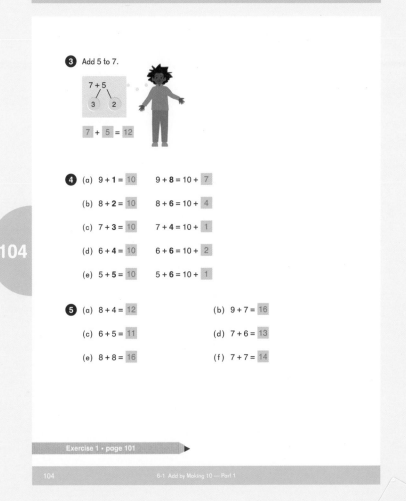

❸ Add 5 to 7.

7 + 5

3   2

7 + 5 = 12

❹ (a) $9 + 1 =$ 10 $9 + 8 = 10 +$ 7

(b) $8 + 2 =$ 10 $8 + 6 = 10 +$ 4

(c) $7 + 3 =$ 10 $7 + 4 = 10 +$ 1

(d) $6 + 4 =$ 10 $6 + 6 = 10 +$ 2

(e) $5 + 5 =$ 10 $5 + 6 = 10 +$ 1

❺ (a) $8 + 4 =$ 12   (b) $9 + 7 =$ 16

(c) $6 + 5 =$ 11   (d) $7 + 6 =$ 13

(e) $8 + 8 =$ 16   (f) $7 + 7 =$ 14

Exercise 1 • page 101

Activities

▲ Teens Target – Addition

Materials: Teens Target – Addition Game Board (BLM) for each player, Number Cards (BLM), Ten-frame Cards (BLM), or playing cards 0 to 9

Similar to the game in **Chapter 5: Lesson 5** on page 132 in this Teacher's Guide, players take turns drawing cards and writing numbers on empty squares on the Teens Target – Addition Game Board (BLM) to make valid equations. If they draw a number that cannot fit in a valid equation, they discard that card. The first player to make five true equations wins.

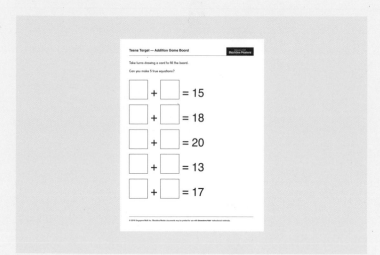

★ Additional Practice

Write additional problems on the board and have students find the missing number. Recall that the equal sign means both expressions are the same. For example, "7 plus 5 is the same as 10 + what number?"

Examples:

$7 + 5 = 10 + \underline{\hspace{1cm}}$

$9 + 6 = 10 + \underline{\hspace{1cm}}$

$9 + \underline{\hspace{1cm}} = 10 + 2$

$8 + \underline{\hspace{1cm}} = 10 + 6$

$\underline{\hspace{1cm}} + 4 = 10 + 2$

$\underline{\hspace{1cm}} + 9 = 10 + 8$

Exercise 1 • page 101

Objective

- Make 10 with 2 one-digit numbers by making a 10 using the second addend.

Lesson Materials

- Linking cubes (10 each of 2 different colors per student) or two-color counters
- Blank Double Ten-frames (BLM), 1 per student

Think

Pose the hanger problem from **Think**. Provide students with linking cubes and a Blank Double Ten-frame (BLM) to help them work through the problem. Have students share the method they used for solving the problem.

Learn

Ask students, "What is different in this problem from the problem with the peppers in the prior lesson?" Possible student answers:

- The 3 is less than the 9.
- The greater number is after the smaller number.
- You need 7 to add to the 3 to make 10.

Provide additional examples with the lesser numbers as the first addend.

Looking at the text, note how Sofia is splitting the number 3 to 2 and 1.

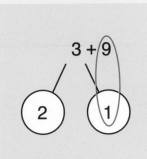

Show this number bond on the board, and emphasize that the 1 and 9 are being combined to make a 10 by circling the numbers. Then add the remaining part of 3, or 2.

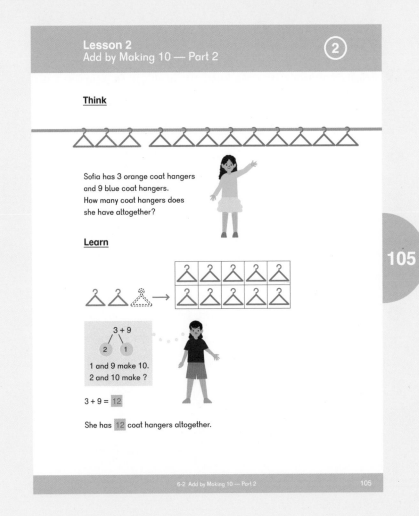

Do

1 — 3 Provide students with Blank Double Ten-frames (BLM) and linking cubes or counters if needed to solve the problems.

4 As in the prior lesson, problems are scaffolded to encourage students to see how many more make 10 in the first column.

In the second column, students may need counters and Blank Double Ten-frames (BLM) to see that 5 + 9 is equal to 10 and 4 more.

5 These problems are designed to be solved by splitting the first addend. Students may find that splitting the second addend for problems (b), (d), and (e) is just as easy for them. Either method is acceptable.

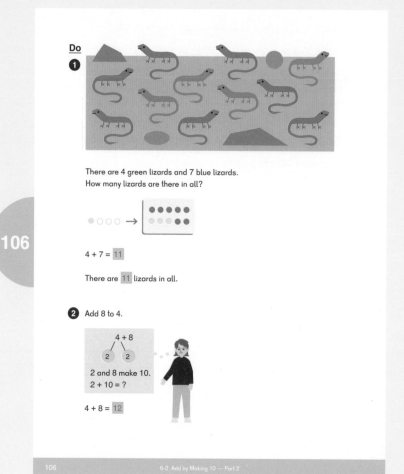

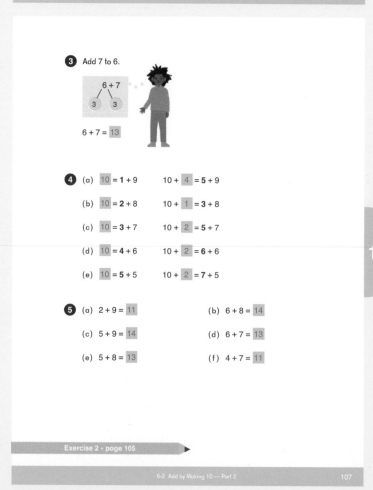

Teacher's Guide 1A Chapter 6   © 2017 Singapore Math Inc.

Activity

▲ Rock, Paper, Scissors, Math!

This is a modification of the game **Rock, Paper, Scissors, Math** that was originally taught in **Chapter 1**: **Lesson 4** on page 15 of this Teacher's Guide.

On the word "Math!," players shoot out some fingers on both hands. The student who says the sum of the fingers first is the winner.

For example, if Player 1 shows 7 fingers and Player 2 shows 6 fingers, the first player to say, "13," wins.

● Struggling students paired together can be encouraged to "match 5s" by putting hands together first to make a 10, then counting the additional fingers.

For example, if Player 1 shows 7 fingers and Player 2 shows 6 fingers, they match up the 5 on each hand and count the 3 more.

Exercise 2 • page 105

Objective

- Add 2 one-digit numbers with sums within 20 by making a 10 with either addend.

Lesson Materials

- Linking cubes or counters, 20 per student (if needed)
- Blank Double Ten-frames (BLM) (if needed)

Think

Pose Sofia's chocolates problem from **Think**. Provide students with linking cubes and a Blank Double Ten-frame (BLM), if needed, to help them work through the problem. Have students share the method they used for solving the problem.

Ask students how they would solve the problem using the strategy of making 10.

Students could work with whiteboards to show how they would decompose the numbers.

Learn

Looking at the text, ask students, "How many solved their problem like Mei? How many solved their problem like Alex?"

Ask students if it matters which way they preferred to solve the problem. Possible student answers:

- 8 is close to 10, so I took 2 from the 7 and made a 10, then I had 5 left. 10 and 5 makes 15.
- 7 is first and 3 more will make a 10, so I took 3 from the 8. Then there's 5 left from the 8. 10 + 5 = 15
- 8 plus 7 is the same as 10 plus 5.

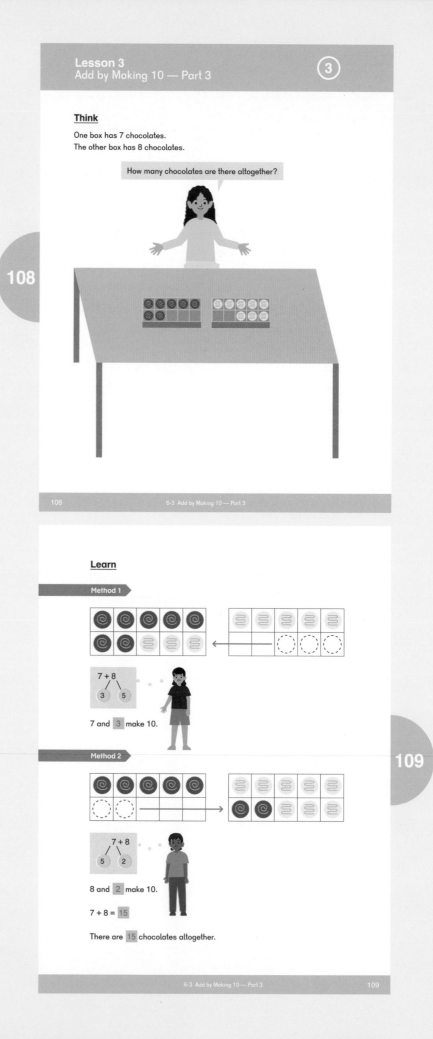

Do

1 — **2** Discuss why one method might be easier than another.

3 Have students show the number bond for the addend they decomposed (or split) to make a ten.

4 — **5** Students should solve the problems by using the strategy of making 10. Allow them to decompose either addend and explain why they chose the addend.

Students may solve the problems in other ways. For example, students may find adding five is easier for them and may solve problems like:

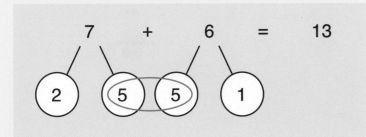

Alternative strategies based on number sense are perfectly acceptable.

To extend, ask students questions using the term "more." For example:

- 8 more than 9 is ?
- 8 more than 9 is 10 and ?

Ask students to write word problems similar to **4**, using numbers from 1 to 9:

- I have 8 toy cars. My friend gives me 9 more cars. How many cars do I have now?

Activity

▲ Addition Face-off

Materials: Deck of Number Cards (BLM) 0 to 10, or regular playing cards

Play in groups of two to four. If using a regular deck of cards, aces are 1 and face cards are 10. Deal cards evenly among players. Each player flips two cards over and calls out the sum. The player with the greatest sum wins and collects all the cards.

If there is a tie, repeat, turning over two more cards to determine the winner.

Exercise 3 • page 109

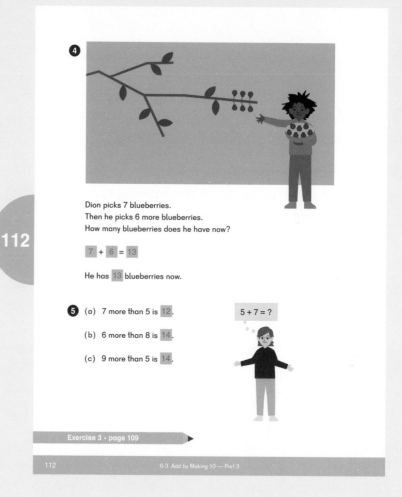

112

④

Dion picks 7 blueberries.
Then he picks 6 more blueberries.
How many blueberries does he have now?

7 + 6 = 13

He has 13 blueberries now.

⑤ (a) 7 more than 5 is 12.

(b) 6 more than 8 is 14.

(c) 9 more than 5 is 14.

5 + 7 = ?

Exercise 3 • page 109

112   6-3 Add by Making 10 — Part 3

Lesson 4 Addition Facts to 20

Objective

- Learn addition facts to 20.

Lesson Materials

- Number Cards (BLM) 11 to 20, 1 set per student
- Spinner 11 – 20 (BLM) or 10-sided die with numbers 11 to 20
- Paper clip for Spinner
- Index cards, 36 per student

Think

Have students use index cards to create their own flash cards similar to the textbook. They should lay the flash cards out and look for patterns. The flash cards can also be used for future practice and games.

Learn

Have students lay out the Number Cards (BLM) and sort their flash cards under each number.

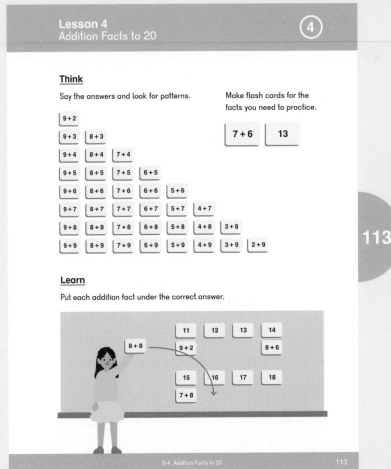

Do

2 Students play in pairs, using Spinner 11 – 20 (BLM).

Activity

▲ Math Fact Jump to 20

Materials: Flash cards made in this lesson, 2 grids created with sidewalk chalk or painter's tape (see below for example)

Played the same as **Math Fact Jump** in **Chapter 4: Lesson 9** on page 103 of this Teacher's Guide, but with addition facts to 20.

One student is the Caller. Two students are the Jumpers and stand on their Home square.

The Caller draws a card and reads the addition equation. Jumpers must jump on the answer.

The first Jumper who misses the correct square becomes the next Caller.

10	19	15
16	17	13
11	14	12
18	20	Home

10	19	15
16	17	13
11	14	12
18	20	Home

Exercise 4 • page 111

Do

1
(a) $8 + 5 = 13$ (b) $6 + 5 = 11$
(c) $4 + 8 = 12$ (d) $7 + 7 = 14$
(e) $9 + 6 = 15$ (f) $5 + 7 = 12$
(g) $6 + 7 = 13$ (h) $9 + 9 = 18$
(i) $8 + 6 = 14$ (j) $7 + 4 = 11$
(k) $6 + 5 = 11$ (l) $8 + 8 = 16$
(m) $9 + 4 = 13$ (n) $5 + 9 = 14$

2 Find addition facts.

$6 + 6 = 12$ $7 + 5 = 12$

6 + 6   7 + 5

9 + 9   9 + 6

5 + 6   9 + 2   6 + 8   7 + 6   7 + 8

Exercise 4 • page 111

114

114   6-4 Addition Facts to 20

Objective

- Practice addition facts to 20 to mastery.

Practice

After students complete the **Practice** in the textbook, have them continue adding numbers to 20 by playing games from this chapter.

Students should be fluent with addition to 20 before moving on to **Chapter 7**: **Subtraction Within 20**.

5 Students can use their fact cards to complete.

7 These questions are an opportunity to discuss that the quantities on each side of an equal sign must have the same value for the equation to be "true."

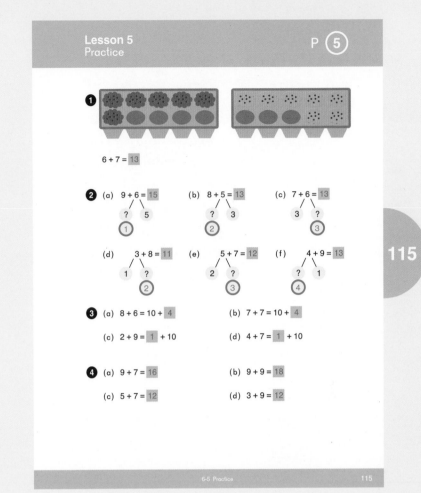

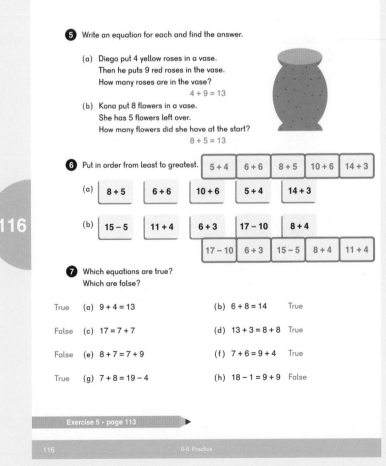

Activity

★ Add 'Em Up Race

6	
8	
9	
4	
10	

Students seeking a challenge will enjoy the race element of this game. Have two to four students go to the board and make a large "T" shape. Randomly call out five numbers and have students write them on the left side of the T as shown on the left.

+ 6	
6	12
8	14
9	15
4	10
10	16

When they have those written, give them a second addend, "Add 6." Students write "+ 6" on the top of the "T," then proceed to solve the problems on the right side as fast as they can, as shown on the left.

Brain Works

★ Lo-shu (Magic Square)

Materials: Number Cards (BLM) 1 to 9

Students try to place the 9 Number Cards (BLM) such that each row, column and diagonal can be added together to get a total of 15.

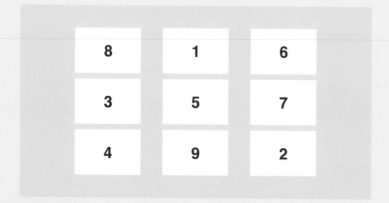

★ KenKen

Easy **KenKen** activities are a great way to practice both addition and logical thinking. Similar to a Sudoku, each row and column contains a number exactly once. The number given in the upper left corner indicates the sum of all the boxes in a cage, or highlighted rectangle, but not the order. Students deduce the answers by solving addition problems.

Some websites allow customized activities to use addition or addition and subtraction only. Search:

- kenkenpuzzle.com
- mathdoku.com
- calcudoku.org

An example **KenKen** board is shown below.

4 +	7 +		3 +
1	4	3	2
	6 +		
3	2	4	1
5 +		1	7 +
2	3	1	4
4	3 +		
4	1	2	3

Exercise 5 · page 113

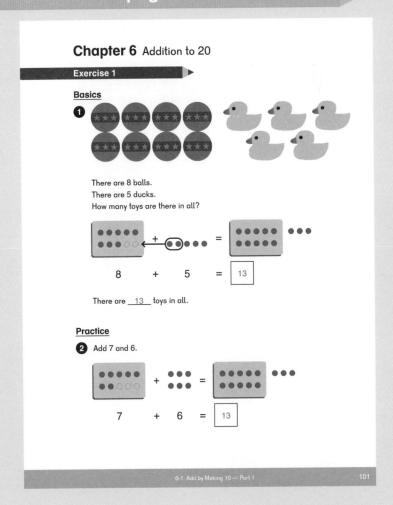

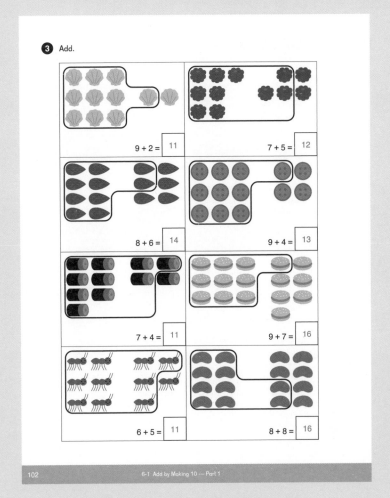

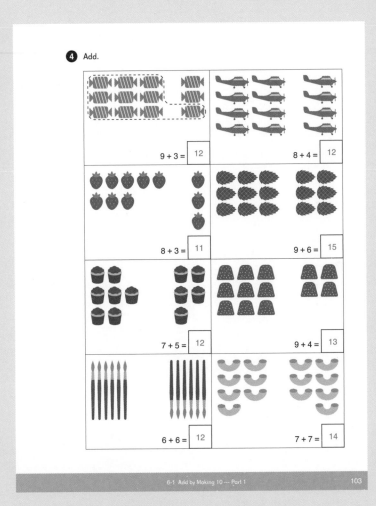

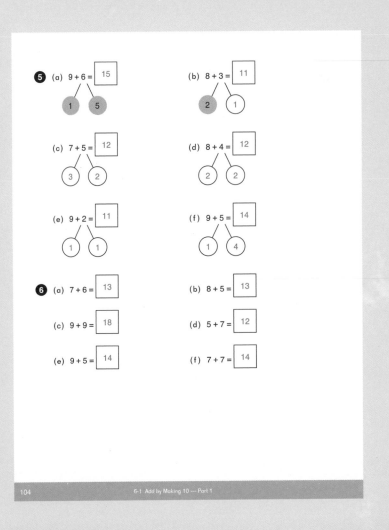

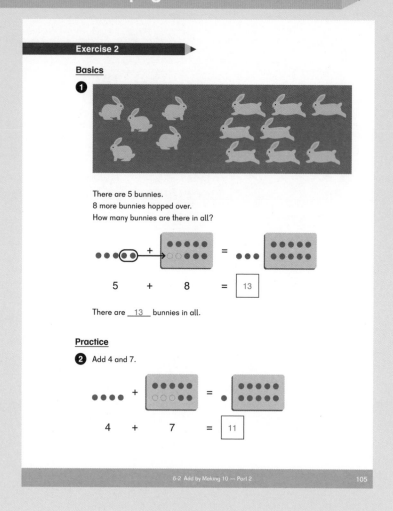

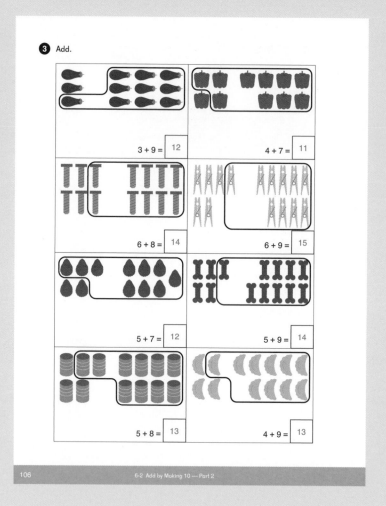

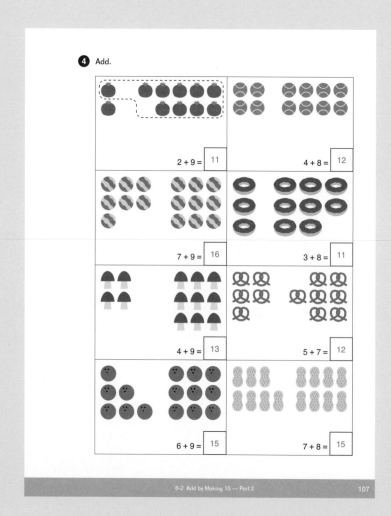

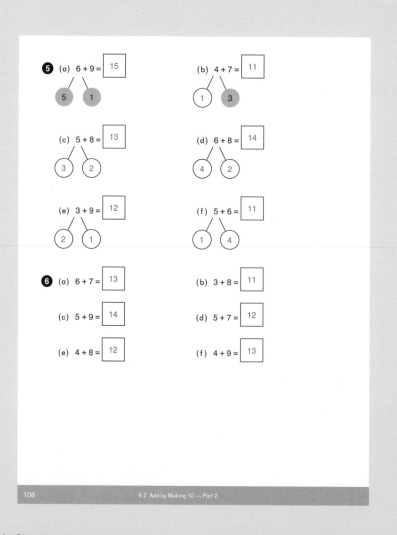

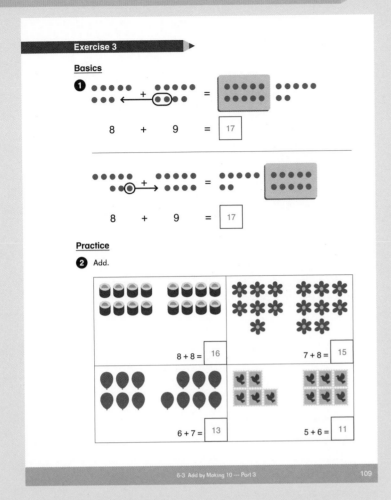

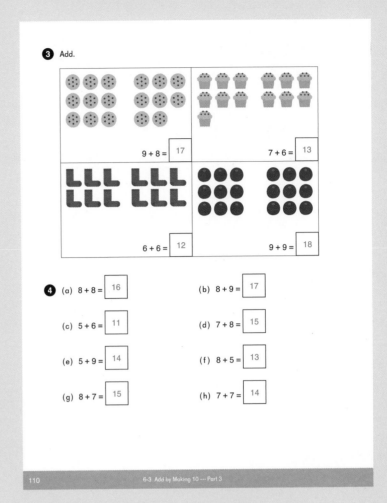

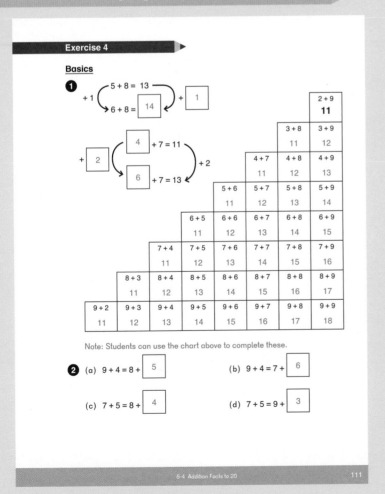

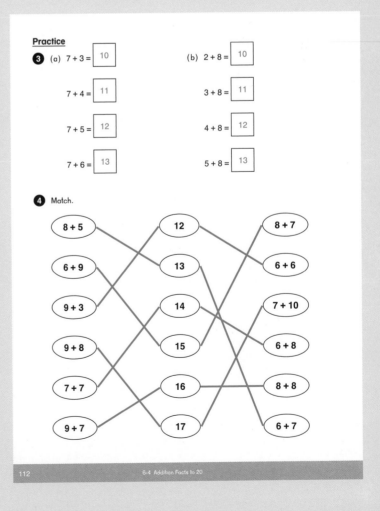

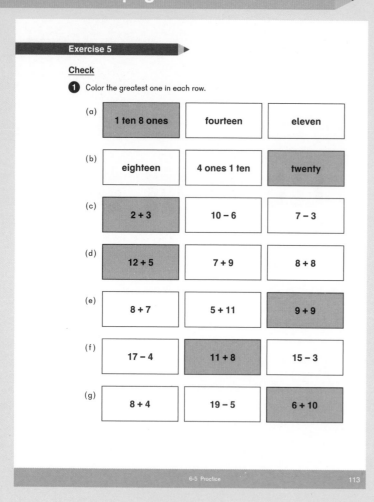

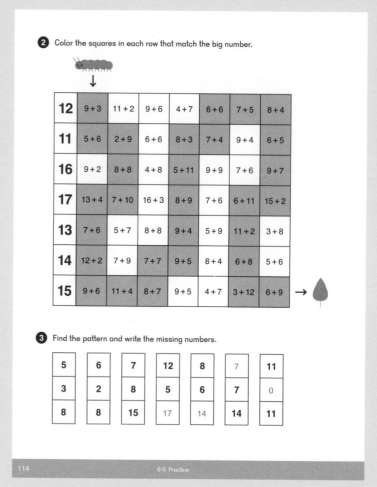

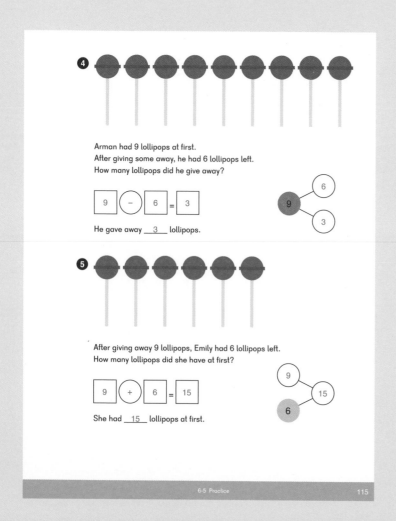

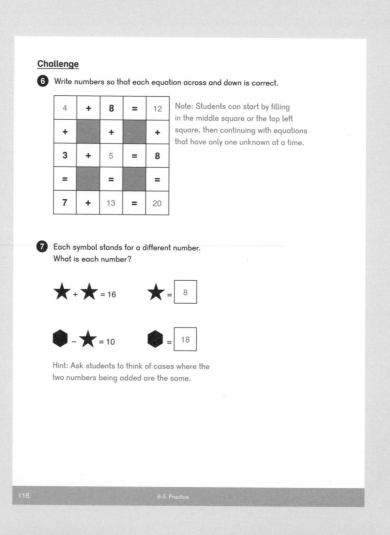

Suggested number of class periods: 6 – 7

Lesson		Page	Resources		Objectives
	Chapter Opener	p. 169	TB:	p. 117	Investigate subtraction within 20.
1	Subtract from 10 — Part 1	p. 170	TB: WB:	p. 118 p. 117	Subtract a one-digit number from a two-digit number.
2	Subtract from 10 — Part 2	p. 173	TB: WB:	p. 121 p. 121	Subtract a one-digit number from a two-digit number.
3	Subtract the Ones First	p. 176	TB: WB:	p. 124 p. 125	Subtract a one-digit number from a two-digit number.
4	Word Problems	p. 180	TB: WB:	p. 131 p. 129	Solve addition and subtraction word problems.
5	Subtraction Facts Within 20	p. 183	TB: WB:	p. 135 p. 131	Learn subtraction facts to 20.
6	Practice	p. 185	TB: WB:	p. 137 p. 135	Practice subtraction facts to 20.
	Workbook Solutions	p. 187			

Chapter 7: Subtraction Within 20 introduces students to subtraction of two numbers that are both less than 20. Students should have mastered the addition and subtraction facts to 10 before this chapter.

In **Chapter 4: Subtraction**, students learned to subtract by counting back 1, 2, or 3. This strategy is not re-taught in this chapter. While some students may continue to use the counting back strategy, the objective of this chapter is to have them learn more efficient strategies that will transfer to larger numbers.

In **Chapter 5: Numbers to 20**, students learned to subtract within 20 without regrouping by subtracting from the ones:

$$18 - 5 = 10 + 3 = 13$$

10   8

In this chapter, students will learn two strategies:

- Subtract from a 10
- Decompose the subtrahend

The first two lessons focus on subtracting from a 10: the student decomposes the minuend into 10 and ones, subtracts the subtrahend from the 10, and adds the remaining ones. This can be a challenging strategy and students will need plenty of practice with concrete materials.

$$13 - 5 = 5 + 3 = 8$$

3   10 $(10 - 5 = 5)$

Lesson 1: Subtract from 10 — Part 1 subtracts only subtrahends of nine and eight. **Lesson 2: Subtract from 10 — Part 2** practices the same skill with subtrahends of 5, 6, and 7. This is best shown with base-10 blocks rather than linking cubes. Decomposing the minuend is a strategy that supports the concept of place value and the concept of equality. This is best shown using linking cubes as the manipulative. These foundational skills will support standard algorithms and algebraic thinking in future years. While students are learning the strategy, they should be encouraged to show how they decomposed with number bonds.

Lesson 3: Subtract the Ones First introduces splitting the subtrahend, sometimes called the "double-subtract method." This strategy is useful when the ones digit of both numbers differ by 1 or 2.

$$13 - 5 =$$

3   2

$$13 - 3 = 10 \quad \text{or} \quad 13 - 3 - 2$$
$$10 - 2 = 8$$

Students may also use known facts from addition or number bonds to solve subtraction problems. If they have mastered $8 + 5 = 13$, they may think of $13 - 5$ as $5 + \_\_\_ = 13$ and simply know the fact.

After **Lesson 5: Subtraction Facts Within 20**, students should be working towards fluency with the facts to 20. They should not be expected to explain every answer. The ultimate goal is automaticity ("I just know that $14 - 8$ is 6."); however, a strategy can be used when facts are not yet automatic.

Students should continue to practice addition and subtraction facts to 20 after their work in this chapter is complete. They should know the facts from memory prior to **Dimensions Math® 1B Chapter 12: Numbers to 40.**

Materials

- Dice
- Counters, including two-color counters
- Linking cubes
- Playing cards
- Shopping lists for classroom items (create per sample on page 183)
- Regular paper cut into quarters
- Index cards or construction paper
- Game board
- Whiteboards

Note: Materials for Activities will be listed in detail in each lesson.

Blackline Masters

- Blank Double Ten-frames
- Teens Target — Subtraction Game Board
- Number Cards
- Subtraction Within 20 Fact Cards
- Ones and Tens Tickets
- Equation Symbol Cards
- Adding Zero Alligator Cards
- Subtracting Zero Alligator Cards

Storybooks

- *The Mission of Addition* by Brian P. Cleary
- *Mission: Addition* by Loreen Leedy
- *12 Ways to Get to 11* by Eve Merriam
- *What's New at the Zoo?* by Suzanne Slade
- *Where's Waldo?* Series by Martin Handford
- *Where is Moshi?* Series by Moshi Moshi Kawaii
- *I Spy* Series by Jean Marzollo and Walter Wick
- *Can You See What I See?* Series by Walter Wick
- *Hey Seymour!* Series by Walter Wick

Letters Home

- Chapter 7 Letter

Notes

Chapter Opener

Objective

- Investigate subtraction within 20.

This **Chapter Opener** is an informal exploration of subtraction within 20.

Students may count objects and tell subtraction stories:

- There are 15 seagulls and 7 are flying away.
- I see 11 cups of juice, kids are drinking 3, and there are 8 left to drink.
- There are 11 shells, 5 are pink and 6 are white.

This lesson may continue straight to **Lesson 1: Subtract from 10 — Part 1**.

Activity

▲ **Ten-frame Wipe-out**

Materials: 1 Die per group or player, 1 Blank Double Ten-frame (BLM) per player, 20 counters per player

This game works best with two to four players. Players begin by putting a counter in each square of both ten-frames on their Blank Double Ten-frame (BLM). Player 1 rolls a die, and subtracts that many counters from one of their ten-frames. Player 2 rolls and does the same. Play continues.

Players must "wipe" one ten-frame first. If a player has 14 and rolls a 5, her turn is over. She must roll down to an exact 10 first, then to an exact 0. If her roll requires her to subtract more counters from her ten-frame than the number of counters she has remaining, her turn is over.

Chapter 7

Subtraction Within 20

14 pretzels
- 6 on plate
- 8 in bag

11 cups
- 3 empty
- 8 full

11 shells
- 5 pink
- 6 white

15 seagulls
- 8 standing
- 7 flying

117

117

Lesson 1 Subtract from 10 — Part 1

Objective

- Subtract a one-digit number from a two-digit number.

Lesson Materials

- Two-color counters, 20 per student
- Blank Double Ten-frames (BLM), 1 per student

Think

Pose the **Think** problem. Students can use counters and Blank Double Ten-frames (BLM) to represent the crackers.

Have students share how they solved the problem. Possible student answers:

- I took away 9 counters (counted back).
- I counted on from 9.
- I used the ten-frames.
- I knew 9 + 4 was 13.

Learn

Have students begin by representing the crackers on the ten-frames as 10 and 3.

Emphasize that if students take away 9 on their first ten-frame, it would be easy to see that there is only 1 square without a counter. Next, count all of the remaining counters, and 1 plus 3 make 4. A student might say, "We take 9 away from the 10, that leaves 1. 1 and 3 make 4."

Provide additional examples with a whole of 11 through 15 and subtracting 9. Follow with subtracting 8. Reinforce the language of taking 9 or 8 away from the 10. For example, "For 15 − 8, we take 8 away from the 10, that leaves 2. 2 and 5 make 7."

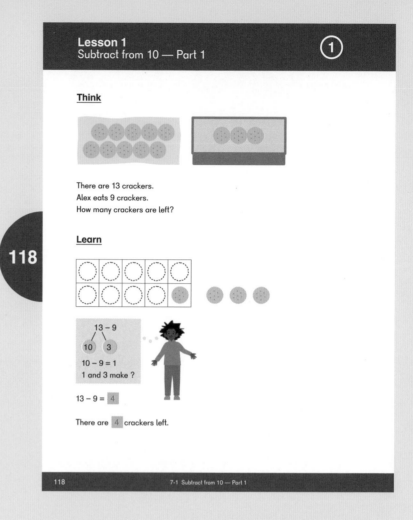

118

Have students discuss how Dion is splitting the number 13 to 10 and 3.

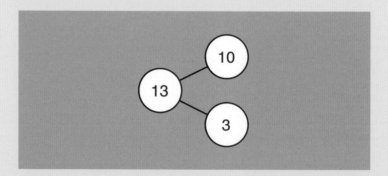

Students may orient their number bonds in any way that helps them remember to subtract from the 10. Teachers should model a variety of orientations for the number bond to support this variability.

Do

While working through the **Do** problems, teachers should model "subtract from 10":

- 13 is 10 and 3, take 9 from the 10, that leaves 1. $1 + 3 = 4$
- $13 - 9$ is the same as $10 + 3$ minus 9. That's the same as $10 - 9 + 3$.

❶ — ❷ Have students show the whole on a Blank Double Ten-frame (BLM) and subtract from the 10.

❹ Problems are scaffolded to encourage students to recall their facts from 10 in the first column. Students should show their solutions on a whiteboard, such as, "10 minus 9 is 1. $14 - 9$ is the same as 10 minus 9 plus 4. 10 minus 9 is ___? So $14 - 9$ is ___?"

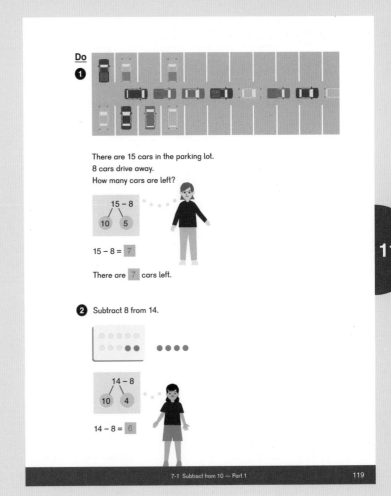

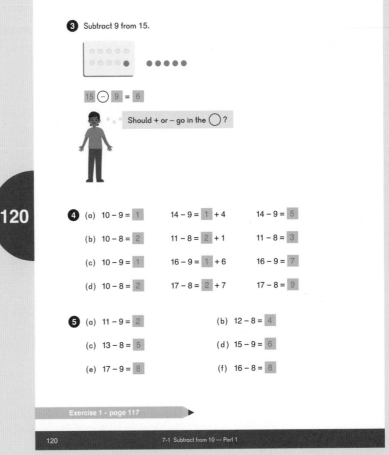

Activities

▲ Teens Target – Subtraction

Materials: 1 Teens Target – Subtraction Game Board (BLM) per player, pile of playing cards 0 to 7, pile of playing cards 8 and 9

Similar to the game in **Chapter 6: Lesson 1** on page 150 of this Teacher's Guide, players take turns drawing cards and writing the numbers on empty squares on their Teens Target – Subtraction Game Board (BLM) to make valid equations. If they draw a number that cannot fit in a valid equation, they discard that card. The first player to make five true equations wins.

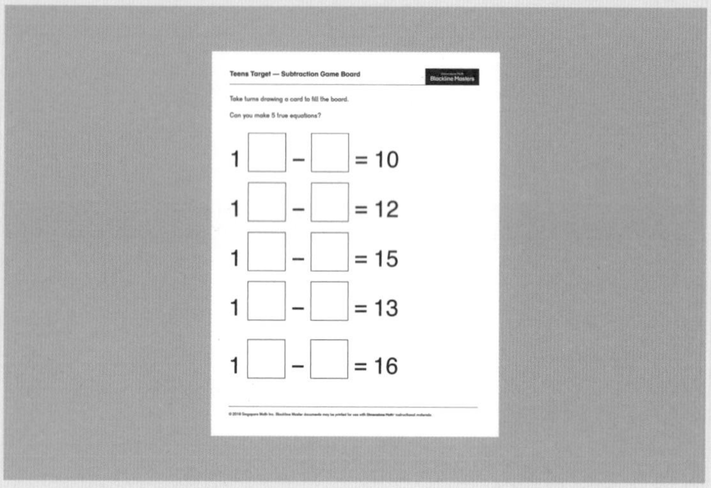

★ Additional Practice

Write additional problems on the board and have students find the missing number.

Examples:

$17 - 9 = 1 + \underline{}$ $16 - \underline{} = 2 + 6$

$16 - 7 = 3 + \underline{}$ $\underline{} - 4 = 10 - 1$

$12 - \underline{} = 3 + 2$ $\underline{} - 5 = 10 - 3$

Exercise 1 • page 117

Teacher's Guide 1A Chapter 7

Lesson 2 Subtract from 10 — Part 2

Objective

- Subtract a one-digit number from a two-digit number.

Lesson Materials

- Linking cubes, 10 each of 2 different colors per student
- Blank Double Ten-frames (BLM)
- Counters
- 12 play fish or other classroom objects

Think

Pose the **Think** problem and provide students with linking cubes and Blank Double Ten-frames (BLM). Have students share their strategies for solving the problem.

Ask students what is similar and what is different in this problem compared to the problem with the crackers in the prior lesson. Possible student answers:

- There are trout instead of crackers.
- I can take 7 from the 10.
- Before we took away only 8 or 9, now we're taking away 7.

Learn

Have students make 10 with one color of cubes and 2 with the other color. Have them take the 7 from the 10. They can then add the 3 and 2 to get 5.

Looking at the text, note how Dion is splitting the number 12 to 10 and 2, and subtracting from the 10.

Provide additional examples subtracting 6 and 7 from wholes of 12 through 15. Have the students use the cubes if needed, and show the number bonds on their whiteboards.

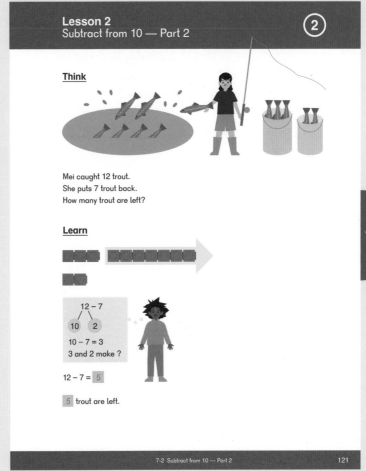

Lesson 2
Subtract from 10 — Part 2 ②

Think

Mei caught 12 trout.
She puts 7 trout back.
How many trout are left?

Learn

12 − 7

10   2

10 − 7 = 3
3 and 2 make ?

12 − 7 = 5

5 trout are left.

7-2 Subtract from 10 — Part 2   121

Do

Provide students with Blank Double Ten-frames (BLM) and counters if needed to solve the problems.

Students may orient their number bonds in any way that helps them remember to subtract from the 10.

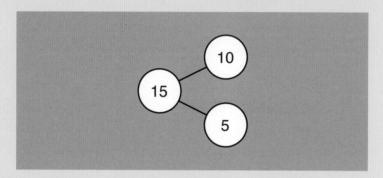

❹ — ❺ Have students discuss their thinking.

Using ❹ (a) as an example, a student might respond, "10 − 7 is 3. 14 − 7 is the same as 10 − 7 plus 4. So 14 − 7 is 3 + 4."

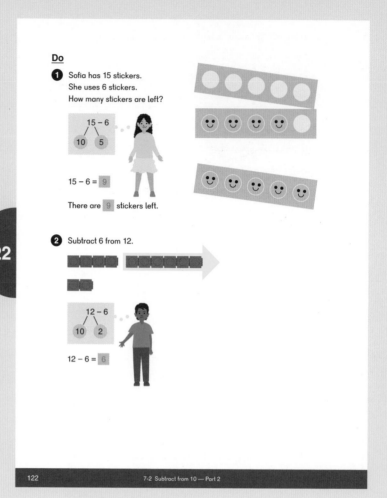

7-2 Subtract from 10 — Part 2

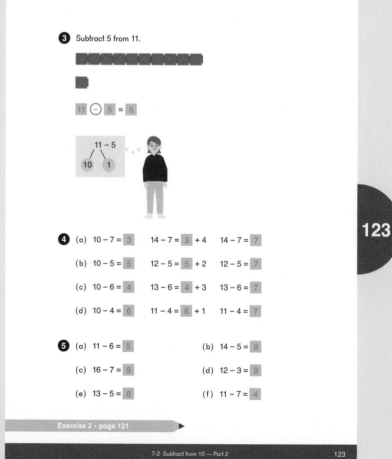

❸ Subtract 5 from 11.

11 ⊖ 5 = 6

11 − 5
10   1

123

❹ (a) 10 − 7 = 3   14 − 7 = 3 + 4   14 − 7 = 7

  (b) 10 − 5 = 5   12 − 5 = 5 + 2   12 − 5 = 7

  (c) 10 − 6 = 4   13 − 6 = 4 + 3   13 − 6 = 7

  (d) 10 − 4 = 6   11 − 4 = 6 + 1   11 − 4 = 7

❺ (a) 11 − 6 = 5   (b) 14 − 5 = 9

  (c) 16 − 7 = 9   (d) 12 − 3 = 9

  (e) 13 − 5 = 8   (f) 11 − 7 = 4

Exercise 2 · page 121

7-2 Subtract from 10 — Part 2   123

Activities

● Match

Materials: Number Cards (BLM) 0 to 10, Subtraction Within 20 Fact Cards (BLM)

Copy each set of cards onto a different color paper.

Lay cards out faceup in an array. Students choose one card of one color and one of another color to find a match.

Have students match a Subtraction Within 20 Fact Card (BLM) to the Number Card (BLM) for the correct answer.

Example:

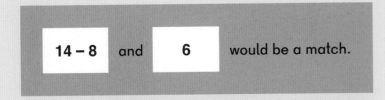

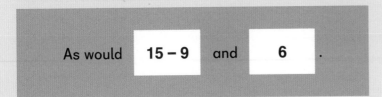

When a match is found, replace the empty spot with new cards. The player with the most matches wins.

▲ Memory

Materials: Number Cards (BLM) 0 to 10, Subtraction Within 20 Fact Cards (BLM)

Play using the same rules as **Match**, but set the cards out facedown in an array.

▲ Shopping Spree

Materials: Ones and Tens Tickets (BLM) (copy on different colors), items in classroom labeled with cost of 5 to 9 tickets, shopping list of items for purchase (see sample below)

Pair students and provide a shopping list for each. Students walk around the room taking turns being the Buyer and the Seller. The Buyer starts with:

To make a purchase, the Buyer hands the Seller his 10-ticket note. The Seller subtracts the cost of the item from 10 and gives the Buyer the difference in 1-ticket notes. The Buyer counts up how many tickets he has now and records the equation on his shopping list.

Exercise 2 • page 121

Lesson 3 Subtract the Ones First

Objective

- Subtract a one-digit number from a two-digit number.

Lesson Materials

- Counters
- Blank Double Ten-frames (BLM)

Think

Pose the **Think** problem and provide students with linking cubes and Blank Double Ten-frames (BLM). Have students share their strategies for solving the problem.

Ask students how they would solve the problem. Method 1 (shown on textbook page 125) is the method students have been working with in prior lessons.

Students could work with whiteboards to show how they would represent the numbers with number bonds. Students can use counters and ten-frames to represent the raspberries if needed.

Students should share and discuss their solutions.

Learn

Have students discuss the solutions that Dion and Alex used and compare them to their own strategy for solving the **Think** problem.

Ask students, "Do you get the same result using either strategy?" Have them share if one method is easier for them than the other.

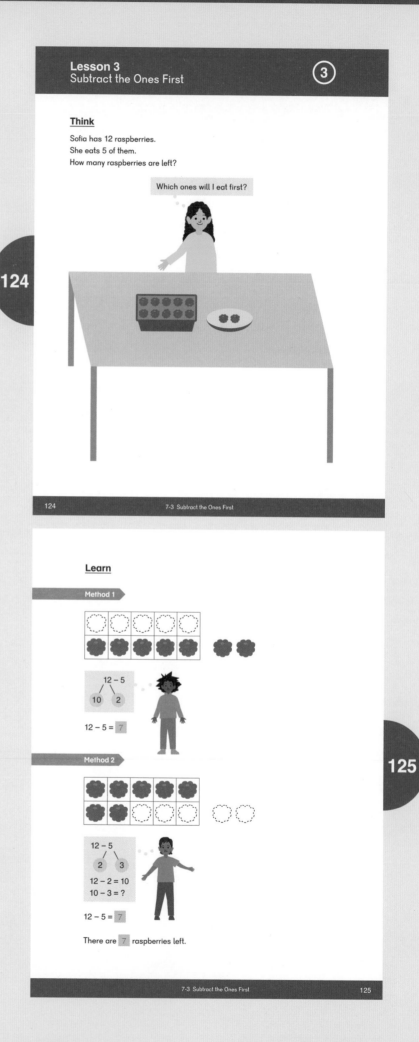

Do

Students should see that these are two ways to solve the same problem, and that these are not two different problems.

❶ Students can use counters and Blank Double Ten-frames (BLM) to work through the problem. Ask students which method is easier for them, Dion's or Emma's.

❷ Ask students if they prefer Mei's method or Sofia's method for this problem. Ask if they used the same method as the prior problem, 13 − 7, or if they used a different method.

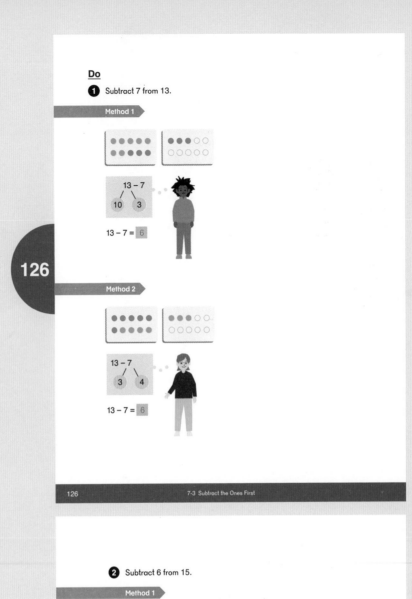

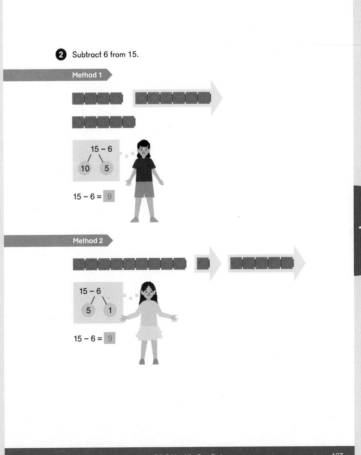

③ — ④ Continue to compare strategies.

⑤ — ⑧ Allow students to use either strategy to solve the problems and have them share how they solved them.

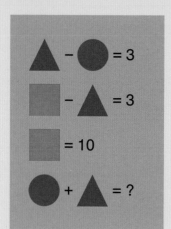

To extend, give students the challenging problem as shown to the left.

To extend the **Do** questions further:

1. Ask students subtraction questions using the term "more," for example:

   - 15 is how many more than 9?
   - 12 is 9 more than what number?

2. Ask students to write word problems such as, "I have 12 toy cars. I have 4 more than my friend. How many cars does my friend have?"

Note: "More" can be used in both addition and subtraction problems.

Activities

▲ Number Stories from Pictures

Materials: Storybooks or children's magazines

Look for storybooks or children's magazines to make number stories within 20 from pictures.

Popular series include:

- *Where's Waldo?* Series by Martin Handford
- *Where is Moshi?* Series by Moshi Moshi Kawaii
- *I Spy* Series by Jean Marzollo and Walter Wick
- *Can You See What I See?* Series by Walter Wick
- *Hey Seymour!* Series by Walter Wick

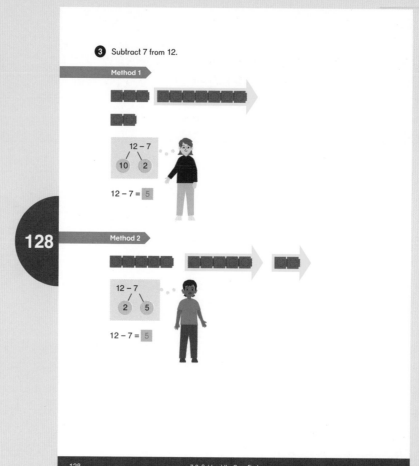

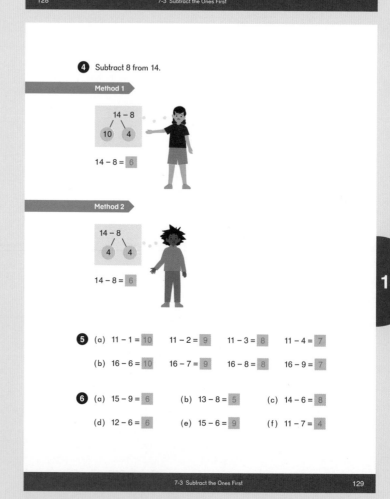

▲ Takeover!

Materials: A game board and markers from the classroom, Subtraction Within 20 Fact Cards (BLM)

This is the same game from **Chapter 4: Lesson 10** on page 104 of this Teacher's Guide.

Most board games require a roll of the dice to determine how many squares to move. In **Takeover!**, players use a deck of Subtraction Within 20 Fact Cards (BLM) to move.

On each turn, the player draws a subtraction fact card and figures out the answer. For example, if a player draws 17 − 9, she moves forward 8 spaces.

To extend, have students create their own board games with special spaces.

★ Subtraction Race

Students seeking a challenge will enjoy the race element of this game.

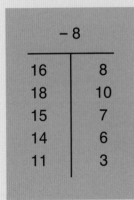

16	
18	
15	
14	
11	

Have two to four students go to the board and make a large "T" shape. Randomly call out five numbers greater than 10 and have students write them on the left side of the "T."

− 8	
16	8
18	10
15	7
14	6
11	3

When they have those written, give them a subtrahend, such as, "Subtract 8."

Students write "− 8" on the top of the "T," then proceed to solve the problems on the right side as fast as they can.

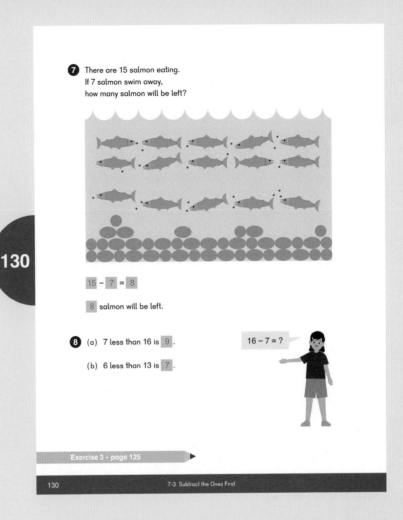

7 There are 15 salmon eating. If 7 salmon swim away, how many salmon will be left?

15 − 7 = 8

8 salmon will be left.

8 (a) 7 less than 16 is 9 .

(b) 6 less than 13 is 7 .

16 − 7 = ?

Exercise 3 • page 125

130   7-3 Subtract the Ones First

130

Exercise 3 • page 125

Lesson 4 Word Problems

Objective

- Solve addition and subtraction word problems.

Lesson Materials

- Linking cubes
- 11 avocados or classroom objects

Note: In this lesson, students use the number bonds to help determine the operation needed to find the missing quantity.

Think

Pose the **Think** problem and provide students with linking cubes. Have students share their strategies for solving the avocado problems.

Ask students what the number bond for 11, 6, and 5 looks like. Ask what equations they can find from the picture or objects.

Learn

Have students look at page 131. Read the two stories aloud and have students compare and discuss the two different stories and equations. Students can use linking cubes to find the solutions if needed.

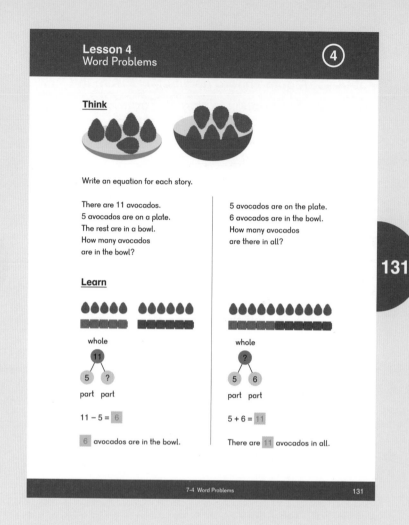

Lesson 4
Word Problems ④

Think

Write an equation for each story.

There are 11 avocados.
5 avocados are on a plate.
The rest are in a bowl.
How many avocados
are in the bowl?

5 avocados are on the plate.
6 avocados are in the bowl.
How many avocados
are there in all?

Learn

whole
11
5 ?
part part

11 − 5 = 6

6 avocados are in the bowl.

whole
?
5 6
part part

5 + 6 = 11

There are 11 avocados in all.

7-4 Word Problems 131

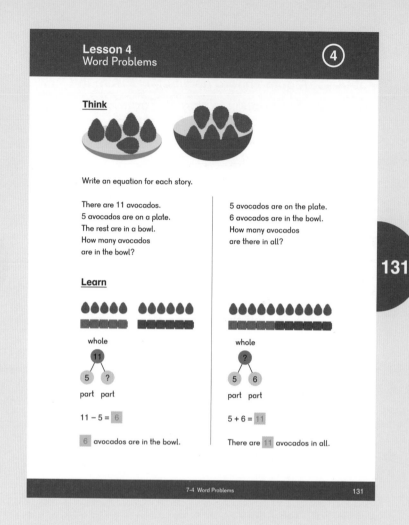

Do

Read the problems aloud to students and have them show their solutions on a whiteboard.

Allow students to use linking cubes to complete the problems if needed.

To extend the **Do** problems, have students write their own word problems.

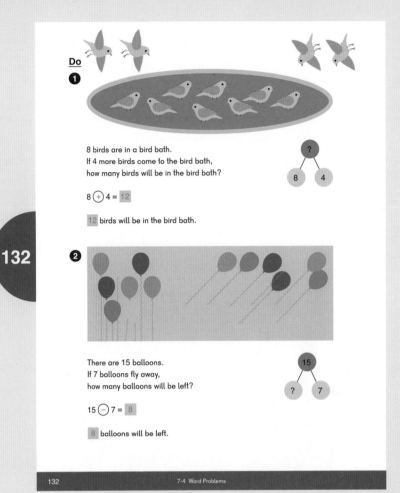

Do

1

8 birds are in a bird bath.
If 4 more birds come to the bird bath,
how many birds will be in the bird bath?

8 ⊕ 4 = 12

12 birds will be in the bird bath.

2

There are 15 balloons.
If 7 balloons fly away,
how many balloons will be left?

15 ⊖ 7 = 8

8 balloons will be left.

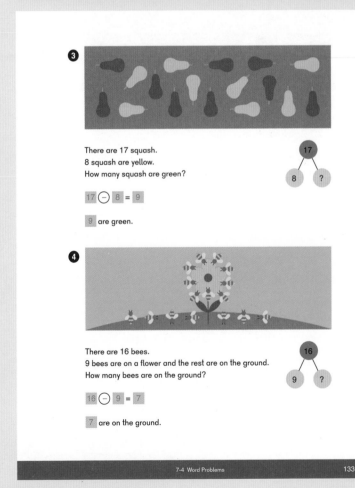

3

There are 17 squash.
8 squash are yellow.
How many squash are green?

17 ⊖ 8 = 9

9 are green.

4

There are 16 bees.
9 bees are on a flower and the rest are on the ground.
How many bees are on the ground?

16 ⊖ 9 = 7

7 are on the ground.

Teacher's Guide 1A Chapter 7

Activities

▲ Word Problem Student Book

Provide each student with different numbers in a number bond format. Have them illustrate pictures from their number bonds similar to the **Think** in the textbook. Collect student stories and put them into a classroom book.

When reading aloud, have students tell or write the equations that could go along with the pictures.

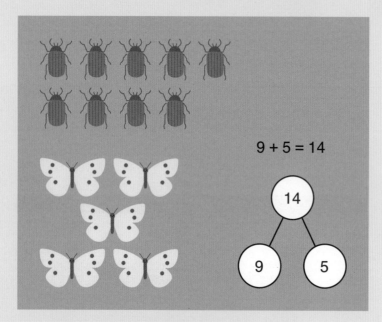

$9 + 5 = 14$

▲ Word Problem Trading Cards

Materials: Regular paper cut into quarters for trading cards

Provide students a few pieces of paper. Have them illustrate a number bond similar to **Think** in the textbook. Provide students with equations, or let them create their own.

Have students write the addition and subtraction equations on the back. Students can trade their cards with classmates and see if their classmates can figure out the equations for the pictures.

Exercise 4 • page 129

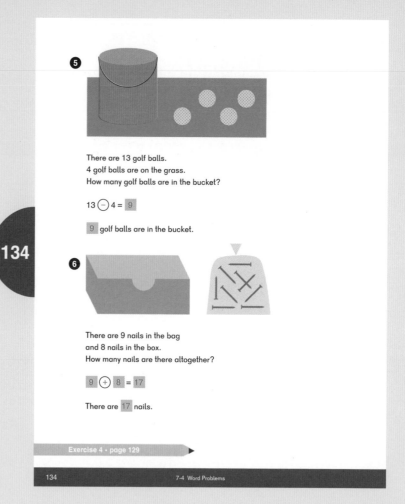

134

5

There are 13 golf balls.
4 golf balls are on the grass.
How many golf balls are in the bucket?

$13 - 4 = 9$

9 golf balls are in the bucket.

6

There are 9 nails in the bag
and 8 nails in the box.
How many nails are there altogether?

$9 + 8 = 17$

There are 17 nails.

Exercise 4 · page 129

134   7-4 Word Problems

Lesson 5 Subtraction Facts Within 20

Objective

- Learn subtraction facts to 20.

Lesson Materials

- Number Cards (BLM) 11 to 20, 1 set per student

Think

Have students use index cards to create their own flash cards similar to the textbook. They should lay the flash cards out and look for patterns. The flash cards can also be used for future practice and games.

Learn

Have students lay out the Number Cards (BLM) 11 to 20 and sort the Subtraction Facts Within 20 flash cards they just created under each number.

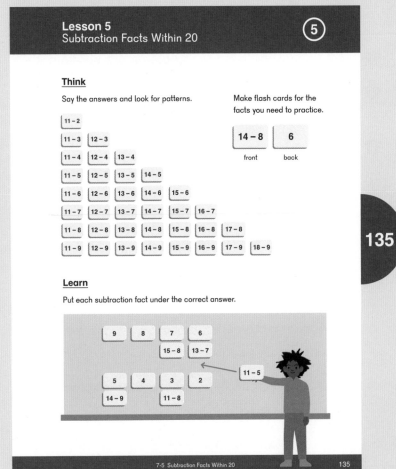

Lesson 5
Subtraction Facts Within 20 ⑤

Think

Say the answers and look for patterns.

11 – 2
11 – 3 12 – 3
11 – 4 12 – 4 13 – 4
11 – 5 12 – 5 13 – 5 14 – 5
11 – 6 12 – 6 13 – 6 14 – 6 15 – 6
11 – 7 12 – 7 13 – 7 14 – 7 15 – 7 16 – 7
11 – 8 12 – 8 13 – 8 14 – 8 15 – 8 16 – 8 17 – 8
11 – 9 12 – 9 13 – 9 14 – 9 15 – 9 16 – 9 17 – 9 18 – 9

Make flash cards for the facts you need to practice.

14 – 8   6
front   back

Learn

Put each subtraction fact under the correct answer.

| 9 | 8 | 7 | 6 |
| | | 15 – 8 | 13 – 7 |

11 – 5

| 5 | 4 | 3 | 2 |
| 14 – 9 | | 11 – 8 | |

135

7-5 Subtraction Facts Within 20   135

Do

❶ — ❷ As in prior lessons, problems are scaffolded.

❸ Students should not have to model the problems at this point.

❹ Students play the game in pairs.

Brain Works

★ **Subtraction Equations**

Materials: Several sets of Number Cards (BLM) 0 to 10, Equation Symbol Cards (BLM)

Ask students, "Using all these cards, can you make two subtraction equations?"

Students may need to try different equations to solve the problem correctly:

$12 - 9 = 3$ or $12 - 3 = 9$

$15 - 7 = 8$ or $15 - 8 = 7$

Students can record or write number stories to go with the equations.

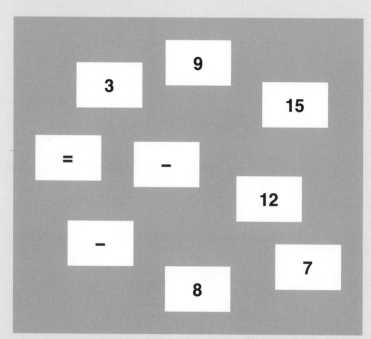

Exercise 5 · page 131

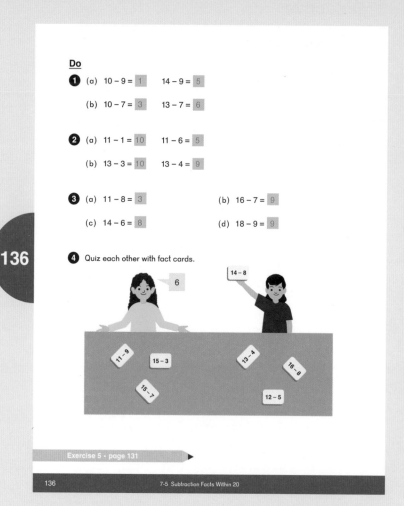

Lesson 6 Practice

Objective

- Practice subtraction facts to 20.

Practice

Continue using the activities from **Chapter 6: Addition to 20** and **Chapter 7: Subtraction Within 20** to practice addition and subtraction within 20. Students should be fluent with these facts prior to **Dimensions Math® 1B Chapter 12: Numbers to 40**.

5 Students can use their fact cards to complete this question.

6 These questions are a great opportunity to discuss that the quantities on both sides of an equal sign must have the same value for the equation to be true.

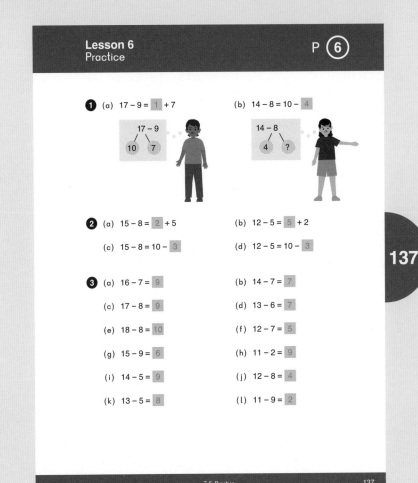

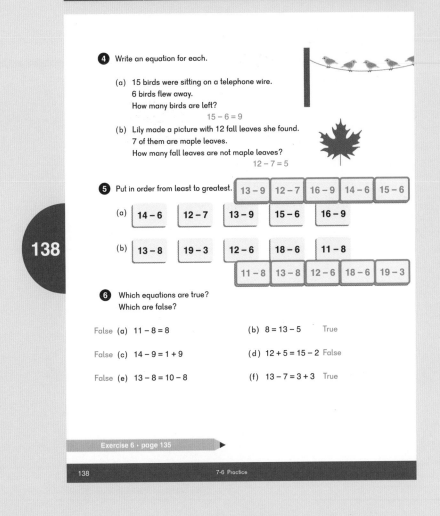

Activities

▲ Salute!

Materials: Deck of cards with face cards removed

Salute! is played with three students. The Caller shuffles, then deals out the deck to two players.

When the Caller says, "Salute!" the other players place the top cards from their piles on their foreheads. The two players can see each other's cards, but not their own.

The Caller tells the players the sum of the two numbers on their cards. (Think of the three players as a number bond with one of the addends missing.)

The player who says her missing part first is the winner. Winners can collect the two cards or players can play through their pile or take turns being the Caller.

The sum is 13.

▲ Alligator! Alligator! Alligator!

Materials: Adding Zero Alligator Cards (BLM), Subtracting Zero Alligator Cards (BLM), or Alligator Cards (BLM) and any other deck of flash cards

Play with Adding Zero Alligator Cards (BLM) or Subtracting Zero Alligator Cards (BLM) to quiz students on their addition and subtraction to 20 facts. You can also add three Alligator Cards (BLM) to any deck of flash cards for practice.

Brain Works

★ Multi-Step Problems

Materials: 20 linking cubes

Provide students multi-step problems and linking cubes to work through the problems.

Two-step: Ava has 2 more markers than Eli. They have 20 markers in all. How many markers does Ava have? (Ava has 11 markers.)

Three-step: Jaina has 3 more markers than Ella. Ben has 2 more markers than Jaina. The three friends have 20 markers in all. How many markers does Jaina have? (Jaina has 7 markers.)

◀ **Exercise 6 • page 135**

Looking Ahead

For the **Chapter Opener** in **Chapter 8**, students will need boxes and cylinders to build with. You may want to ask them to bring in empty shoe boxes, tissue boxes, paper towel rolls, cans, etc., so there are plenty of them for students to use.

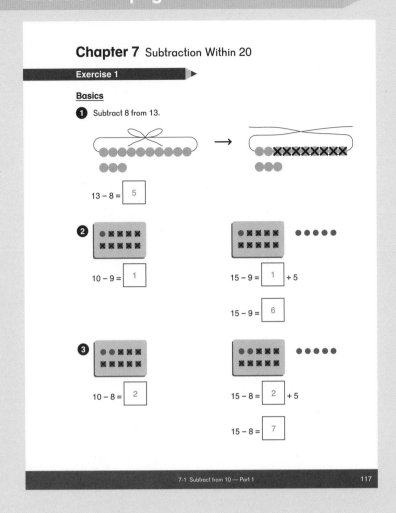

Chapter 7 Subtraction Within 20

Exercise 1

Basics

1 Subtract 8 from 13.

13 − 8 = 5

2

10 − 9 = 1

15 − 9 = 1 + 5

15 − 9 = 6

3

10 − 8 = 2

15 − 8 = 2 + 5

15 − 8 = 7

7-1 Subtract from 10 — Part 1   117

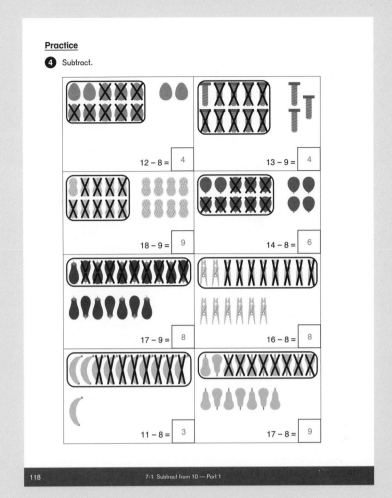

Practice

4 Subtract.

12 − 8 = 4   13 − 9 = 4

18 − 9 = 9   14 − 8 = 6

17 − 9 = 8   16 − 8 = 8

11 − 8 = 3   17 − 8 = 9

118   7-1 Subtract from 10 — Part 1

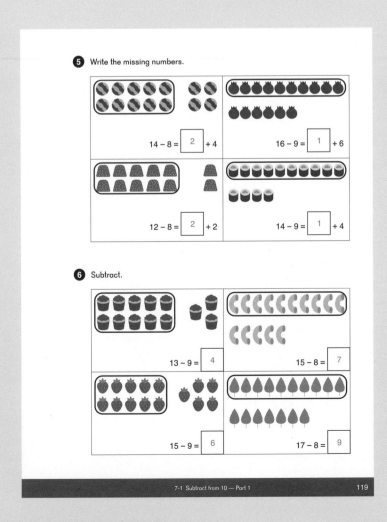

5 Write the missing numbers.

14 − 8 = 2 + 4   16 − 9 = 1 + 6

12 − 8 = 2 + 2   14 − 9 = 1 + 4

6 Subtract.

13 − 9 = 4   15 − 8 = 7

15 − 9 = 6   17 − 8 = 9

7-1 Subtract from 10 — Part 1   119

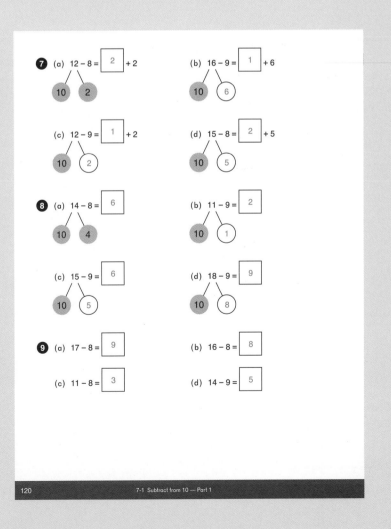

7 (a) 12 − 8 = 2 + 2   (b) 16 − 9 = 1 + 6
        10 2   10 6

(c) 12 − 9 = 1 + 2   (d) 15 − 8 = 2 + 5
    10 2   10 5

8 (a) 14 − 8 = 6   (b) 11 − 9 = 2
        10 4   10 1

(c) 15 − 9 = 6   (d) 18 − 9 = 9
    10 5   10 8

9 (a) 17 − 8 = 9   (b) 16 − 8 = 8

(c) 11 − 8 = 3   (d) 14 − 9 = 5

120   7-1 Subtract from 10 — Part 1

Teacher's Guide 1A Chapter 7

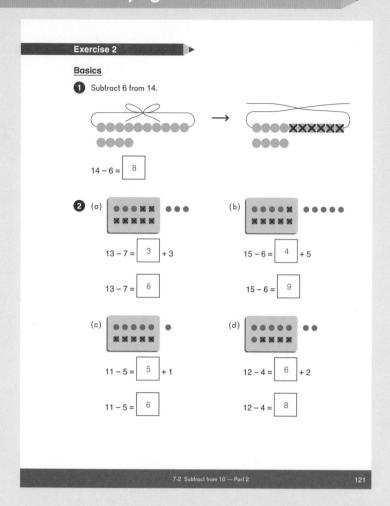

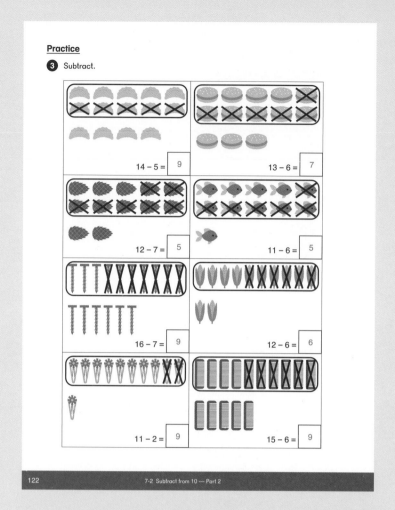

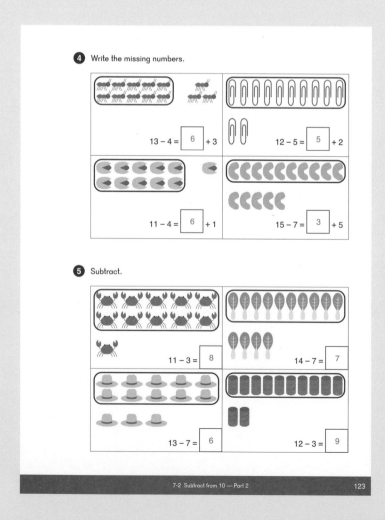

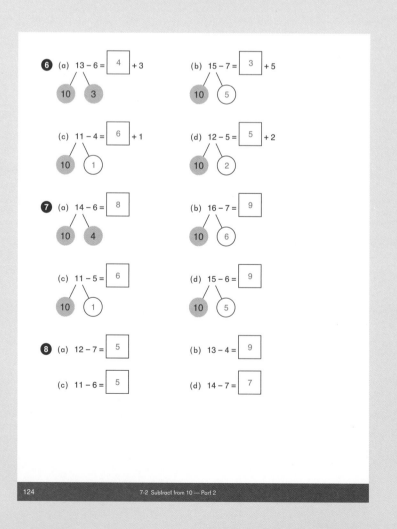

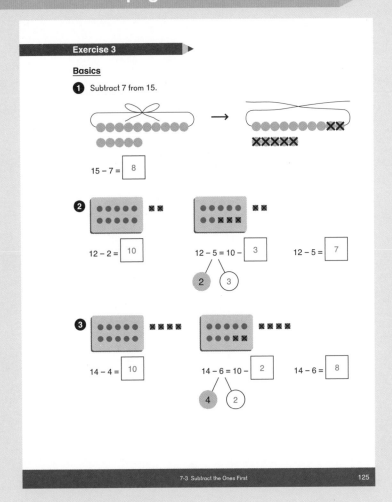

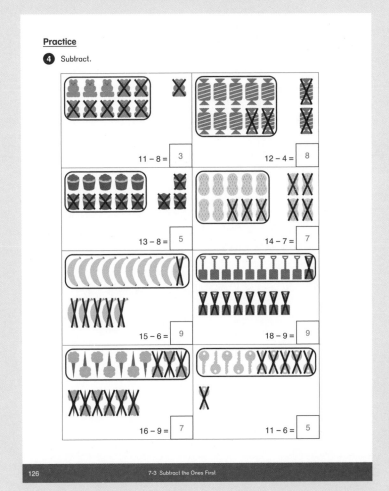

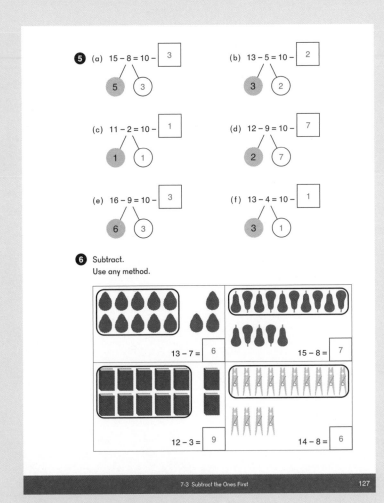

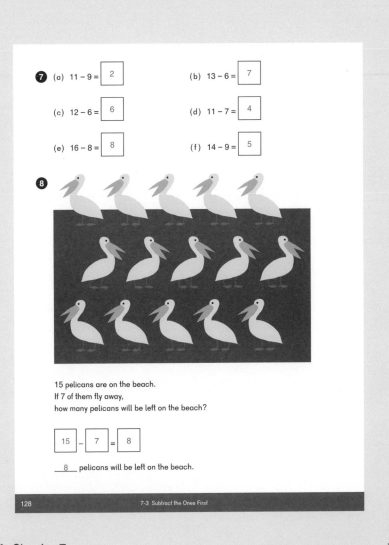

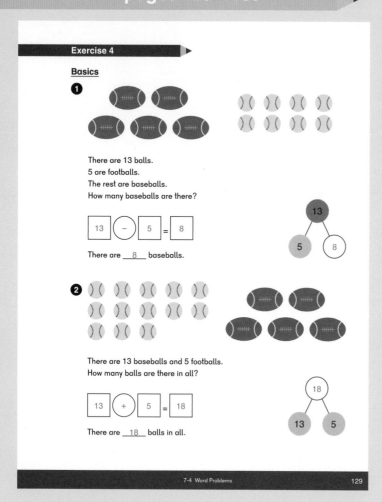

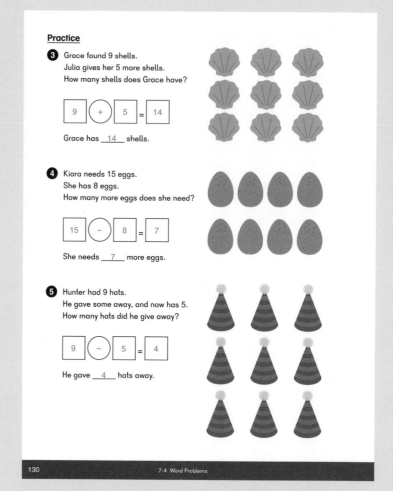

Exercise 5

Basics

1

11 – 2 **9**	11 – 3 8	11 – 4 7	11 – 5 6	11 – 6 5	11 – 7 4	11 – 8 3	11 – 9 2
	12 – 3 9	12 – 4 8	12 – 5 7	12 – 6 6	12 – 7 5	12 – 8 4	12 – 9 3
		13 – 4 9	13 – 5 8	13 – 6 7	13 – 7 6	13 – 8 5	13 – 9 4
			14 – 5 9	14 – 6 8	14 – 7 7	14 – 8 6	14 – 9 5
				15 – 6 9	15 – 7 8	15 – 8 7	15 – 9 6
					16 – 7 9	16 – 8 8	16 – 9 7
						17 – 8 9	17 – 9 8
							18 – 9 9

$11 - 7 = 4$
$11 - 8 = 3$ — −1
$11 - 9 = \boxed{2}$ — $-\boxed{1}$

$11 - 4 = 7$
$12 - 4 = 8$ — +1
$13 - 4 = \boxed{9}$ — $+\boxed{1}$

Note: Students can use the chart above to complete these.

2 (a) $15 - 6 = 14 - \boxed{5}$ (b) $15 - 6 = 13 - \boxed{4}$

(c) $13 - 6 = 14 - \boxed{7}$ (d) $13 - 6 = 15 - \boxed{8}$

Practice

3 (a) $15 - 9 = \boxed{6}$ (b) $16 - 7 = \boxed{9}$

$15 - 8 = \boxed{7}$ $15 - 7 = \boxed{8}$

$15 - 7 = \boxed{8}$ $14 - 7 = \boxed{7}$

$15 - 6 = \boxed{9}$ $13 - 7 = \boxed{6}$

4 Match.

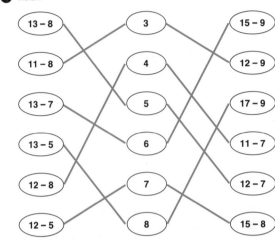

$13 - 8$	3	$15 - 9$
$11 - 8$	4	$12 - 9$
$13 - 7$	5	$17 - 9$
$13 - 5$	6	$11 - 7$
$12 - 8$	7	$12 - 7$
$12 - 5$	8	$15 - 8$

5 Color the squares in each row that match the big number.

7	18 – 9	19 – 2	14 – 7	11 – 4	16 – 9	13 – 6	12 – 5
4	13 – 5	11 – 6	13 – 9	16 – 2	11 – 8	11 – 9	12 – 7
8	17 – 8	17 – 5	12 – 4	16 – 8	11 – 3	14 – 6	15 – 7
2	11 – 6	19 – 4	13 – 8	14 – 2	18 – 3	12 – 4	11 – 9
9	12 – 3	16 – 7	13 – 4	15 – 6	11 – 2	16 – 6	14 – 5
5	11 – 6	19 – 4	19 – 7	13 – 9	14 – 9	11 – 5	12 – 7
6	12 – 6	19 – 3	16 – 4	17 – 9	11 – 5	14 – 8	15 – 9

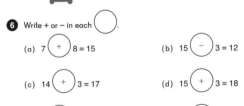

6 Write + or − in each ◯.

(a) $7 \,\boxed{+}\, 8 = 15$ (b) $15 \,\boxed{-}\, 3 = 12$

(c) $14 \,\boxed{+}\, 3 = 17$ (d) $15 \,\boxed{+}\, 3 = 18$

(e) $4 \,\boxed{+}\, 3 = 7$ (f) $15 \,\boxed{+,-}\, 0 = 15$

7 Write the answers.
Color all the boxes with the answer 8 or 12.

19 – 6 13	13 + 6 19	8 + 4 12	17 – 5 12	5 + 7 12	12 – 7 5	18 – 10 8	11 + 1 12
12 – 4 8	10 – 2 8	14 – 6 8	7 + 8 15	9 + 3 12	11 + 5 16	15 – 7 8	15 – 5 10
12 + 0 12	15 – 6 9	7 + 9 16	16 – 9 7	13 – 5 8	9 + 9 18	3 + 5 8	13 – 0 13
17 – 9 8	10 + 2 12	19 – 7 12	6 + 7 13	0 + 12 12	7 + 5 12	11 – 3 8	14 + 2 16
20 – 10 10	10 + 8 18	16 – 8 8	5 + 6 11	18 – 9 9	0 + 14 14	9 + 4 13	4 + 7 11

8 (a) $17 - \boxed{8} = 9$ (b) $11 - \boxed{9} = 2$

(c) $14 + \boxed{3} = 17$ (d) $\boxed{13} - 5 = 8$

(e) $\boxed{8} + 7 = 15$ (f) $9 + \boxed{9} = 18$

Exercise 6

Check

Note: Students may also fill in = signs first in some cases, e.g., 14 = 8 + 6 instead of 14 − 8 = 6.

1 Make as many addition and subtraction equations as you can. 3 are done.

8	+	9	=	17	−	7	=	10		11
−		+		−				−		−
6	−	2	=	4		14	−	5	=	9
=		=		=				=		=
2	+	11	=	13		7	−	5	=	2
										+
3	+	8	=	11	+	6	=	17		2
+				−				−		=
11		17	−	5	=	12	−	8	=	4
=				=				=		
14	−	8	=	6		5	+	9	=	14

2 Cross out any that are greater than 13.

  9 + 2 15 − 6 12 ✕ 3 18 ✕ 4

3

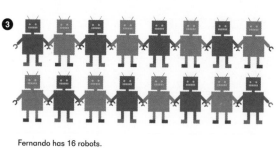

Fernando has 16 robots.
He loses 3 robots.
How many robots does he have now?

 16 ⊖ 3 = 13

He has __13__ robots now.

4

Katherine has a box of 8 paint tubes.
She got another box of 8 paint tubes.
How many paint tubes does she have now?

 8 ⊕ 8 = 16

She has __16__ paint tubes now.

5

Daniel's bag of 8 bananas is full.
He has 5 more bananas.
How many bananas does he have in all?

 8 ⊕ 5 = 13

He has __13__ bananas.

6

Asimah's box of 12 donuts only has 8 donuts.
How many donuts are missing?

 12 ⊖ 8 = 4

__4__ donuts are missing.

Challenge

7 Andrei's box will hold 12 crayons.
It has 9 crayons.
He gets 5 more crayons.
How many crayons will not fit in the box?

There are __2__ extra crayons.
Hint: Students can act the problem out with counters.

8 Write + or − in each ◯.

(a) 9 ⊕ 4 = 16 ⊖ 3   (b) 14 ⊖ 8 = 9 ⊖ 3   Hint: Students can act the problem out with ten-frames.

9 There are 15 dots on 2 ten-frame cards.
How many dots are covered up?

 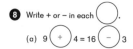 6

10 Each symbol stands for a different number.
What is each number?

 − = 4   = 3   Hint: Suggest that students try to find the value for the shapes in the second equation first.

 + = 14   = 7

Suggested number of class periods: 4 – 5

Lesson		Page	Resources		Objectives
	Chapter Opener	p. 195	TB:	p. 139	Investigate 3D solids and their attributes.
1	Solid and Flat Shapes	p. 196	TB: WB:	p. 140 p. 139	Identify the faces of some 3D solids as circles, rectangles, squares, and triangles.
2	Grouping Shapes	p. 200	TB: WB:	p. 145 p. 143	Group shapes according to size, shape, and color. Create and identify patterns.
3	Making Shapes	p. 203	TB: WB:	p. 149 p. 147	Compose shapes from other shapes.
4	Practice	p. 205	TB: WB:	p. 152 p. 151	Work with shapes.
	Workbook Solutions	p. 207			

Chapter 8: Shapes focuses on hands-on activities that allow students to physically experience solid and flat shapes. In **Dimensions Math® Kindergarten A** students learned to:

- Identify and name three-dimensional (3D) shapes: cube, sphere, cylinder, and cone.
- Identify and name two-dimensional (2D) shapes: square, circle, rectangle, and triangle.
- Use positional words and phrases: above, below, beside, in front of, behind, next to, beneath, inside, and outside.
- Combine shapes to make larger shapes.

In this chapter, students will identify the shapes on the faces of 3D solids. They will also describe and classify shapes based on the number of sides and types of angles, although formal language (such as "right angle") will not be used. Students will describe shapes through patterns, and determine whether they are composed of or can be used to make other shapes.

Attributes that define shapes are edges, sides, angles, and vertices. Shapes are not defined by orientation, color, or size. This chapter allows students to look at all of these attributes in order for them to begin to understand which attributes define a shape.

Students will also look at the attributes of shapes being used in patterns. They will extend patterns defined by shape, orientation, color, and size.

Manipulating shapes by flipping, rotating, and sliding them will help students develop visual-spatial awareness and lay the foundation for later geometry concepts. Students should be allowed plenty of time to investigate shapes.

Materials

- Boxes and cylinders such as shoe boxes, paper towel rolls, etc.
- Pieces of square paper, white and colored
- Black construction paper
- Art paper
- White crayons
- Pattern blocks
- Attribute blocks
- Colored paper
- Yarn or string
- Glue
- Circles cut out of colored paper
- Whiteboards

Note: Materials for Activities will be listed in detail in each lesson.

Blackline Masters

- Shape Sudoku
- Shapes Worksheet
- Tangram Pieces

Storybooks

- *The Shape Of Things* by Dayle Ann Dodds
- *Cubes, Cones, Cylinders, & Spheres* by Tana Hoban
- *Shapes, Shapes, Shapes* by Tana Hoban
- *Three Pigs, One Wolf, Seven Magic Shapes* by Grace Maccarone
- *Shape by Shape* by Suse MacDonald
- *I Spy Shapes in Art* by Lucy Micklethwait
- *Skippyjon Jones: Shape Up* by Judy Schachner
- *Grandfather Tang's Story* by Ann Tompert
- *Mouse Shapes* by Ellen Stoll Walsh

Letters Home

- Chapter 8 Letter

Objective

- Investigate 3D solids and their attributes.

Lesson Materials

- Boxes and cylinders for students to use for building (shoe boxes, tissue boxes, paper towel rolls, cans, etc.)

Ask students to bring 3D solids such as boxes and cans (cylinders) from home. Have them build structures like the friends in the **Chapter Opener**.

Discuss with students the structures they built and the attributes of the 3D solids. For example, you may ask:

- Which ones roll? Which ones slide?
- Which ones look like boxes?
- Which ones have circles on them?

This activity informally explores the attributes of 3D solids to prepare students for **Lesson 1: Solid and Flat Shapes**, where they will identify 2D shapes on 3D solids.

Chapter 8

Shapes

What did they make?
What shapes did they use?

139

139

Lesson 1 Solid and Flat Shapes

Objective

- Identify the faces of some 3D solids as circles, rectangles, squares, and triangles.

Lesson Materials

- 3D solids: cuboids, pyramids, cylinders, rectangular prisms, cubes, and cones for each group of students
- Art paper for each student
- Square pieces of paper, at least 1 for each student

Think

Give students solid shapes and have them trace around the edges on art paper to create a picture.

Students should notice that some solids can be used to draw more than one flat shape. A pyramid, for example, can be used to draw a triangle or square.

Have students share their pictures with the class. They should identify which face on the solid shapes they traced to create their pictures.

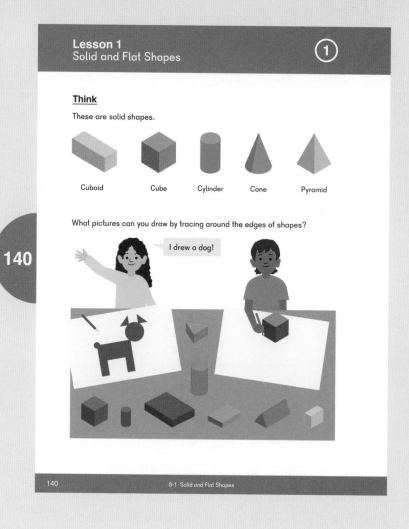

140

Learn

Have students discuss Mei's question. Have them compare the rectangles next to Mei with the squares next to Alex. Discuss Alex's comments about special rectangles.

Note that all squares are rectangles, but not all rectangles are squares.

The term "right angle" does not need to be introduced, but students can visualize what the corners of rectangles and squares look like:

- It looks like the letter L.
- The lines go together but don't cross like an "x."
- The lines are straight, not wavy.

Have students discuss the squares and the rectangles. Students should notice that both rectangles and squares have 4 sides. However, unlike a rectangle, the 4 sides of a square are the same length.

Give students a square piece of paper and have them fold it over diagonally, or overlay and rotate two squares of the same size to demonstrate that the 4 sides of a square are all the same length.

Discuss the shapes and Emma and Dion's thoughts with students. Students should note that some triangles have corners like squares and rectangles and some don't. They should see that a triangle always has 3 sides and 3 corners, regardless of its orientation.

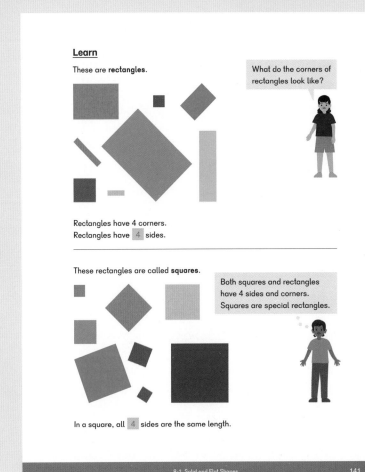

Do

1 Have students go on a shape scavenger hunt around the classroom or school searching for examples of rectangles, squares, triangles, and circles.

3 This question asks students to compare some basic quadrilaterals with rectangles, focusing on the angles. Students do not need to know the terms "parallelogram" or "trapezoid" at this point.

Activities

▲ City at Night

Materials: Black construction paper, white crayons

Using only rectangles, squares, circles, and triangles, have students design and draw the outline of a city at night.

Do

1 Find rectangles, squares, triangles, and circles on objects around you.

8-1 Solid and Flat Shapes   143

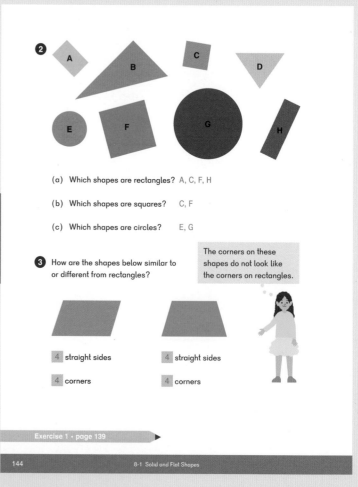

2

A   B   C   D
E   F   G   H

(a) Which shapes are rectangles?   A, C, F, H

(b) Which shapes are squares?   C, F

(c) Which shapes are circles?   E, G

3 How are the shapes below similar to or different from rectangles?

> The corners on these shapes do not look like the corners on rectangles.

4 straight sides   4 straight sides

4 corners   4 corners

Exercise 1 • page 139

144   8-1 Solid and Flat Shapes

▲ Shape Sudoku

Materials: Shape Sudoku (BLM), pattern blocks

Print out either full color or black and white Shape Sudoku (BLM) cards and have students use pattern blocks to complete the game.

Each row, column, and four-square area bordered by a darker line should have one, and only one, of each shape. There may be more than one possible solution on each card.

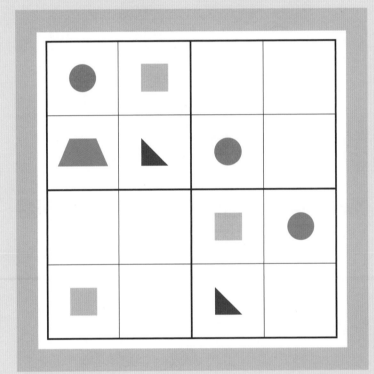

▲ Which One Doesn't Belong?

Show students the pictures below and ask them which shape in each set doesn't belong and why. There could be multiple reasons why each shape may not belong.

A few possible student answers for Set A:

- The rectangle doesn't belong because it doesn't have a corner pointing up.
- The rectangle doesn't belong because it is a different color.
- The top triangle doesn't belong because it doesn't have a special corner.
- The square doesn't belong because it's tilted and the rest are flat.
- The square doesn't belong because the rest have "angle" in their names.

During the discussion, there may be opportunities to correct misconceptions.

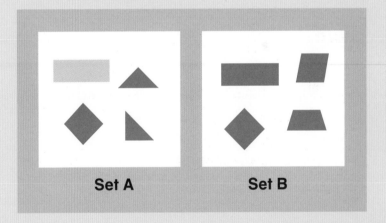

Set A   Set B

Exercise 1 • page 139

Lesson 2 Grouping Shapes

Objectives

- Group shapes according to size, shape, and color.
- Create and identify patterns.

Lesson Materials

- Shapes Worksheet (BLM) or similar colored shapes in different sizes

Think

Have students cut out the set of shapes from the Shapes Worksheet (BLM).

Have students sort their shapes and discuss how they grouped them. Students may have grouped their shapes by:

- Shape
- Color
- Size

Learn

Have students look at the textbook examples on page 146. Ask them how the shapes are grouped. Ask, "How many sides? How many corners?"

Lesson 2
Grouping Shapes ②

Think
Group these shapes in different ways.

145

8-2 Grouping Shapes   145

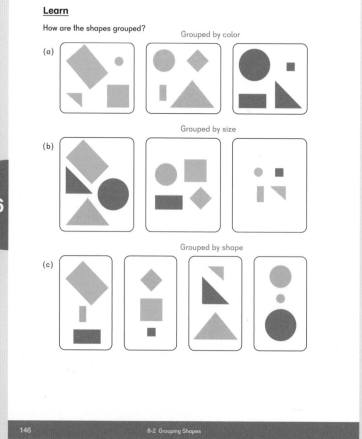

Learn
How are the shapes grouped?

Grouped by color

(a)

Grouped by size

(b)

146

Grouped by shape

(c)

146   8-2 Grouping Shapes

Do

2 Have students describe the next shape in the pattern.

3 Extend by having students draw their own patterns.

Activities

▲ Block Sort

Materials: Attribute blocks

Have students sort and classify attribute blocks. Many attribute blocks sets will have hexagons and other shapes whose names have not been introduced. Remove the shapes if needed, or allow them to provide a challenge for students.

Have students explain how they classified the blocks based on their shapes and sides.

▲ Patterns

Materials: Pattern or attribute blocks

Have students work in pairs. Player 1 should create a pattern using pattern or attribute blocks. Player 2 can find the next three shapes to the pattern.

★ For a greater challenge, Player 1 can create a pattern with missing pieces and Player 2 can find the missing pieces.

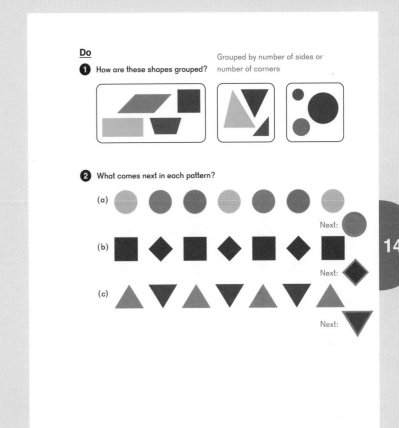

▲ What's My Rule?

Materials: Attribute blocks

Play in small groups of 2 – 5 students. Player 1 starts sorting her shapes, but doesn't reveal the rule or attribute that they she is sorting by.

Other players watch the student and the first one to shout out the rule wins the round.

Students take turns sorting and guessing Player 1's rule.

▲ Pattern Chains

Materials: Colored paper, yarn or string, glue

Have students use colored paper to cut out shapes from the lesson and make a pattern. Once they have a pattern of 5 or 6 shapes, they will need to cut out a match for each of their shapes. Glue them together with the yarn in the middle and display the pattern chains.

The paper chains can be used again in **Chapter 9: Lesson 2** for the **Pattern Chains** activity on page 218 of this Teacher's Guide.

Exercise 2 • page 143 ▶

Lesson 3 Making Shapes

Objective

- Compose shapes from other shapes.

Lesson Materials

- Colored square paper, several pieces per student
- Colored paper, cut into circles

Think

Provide students with squares of paper and have them fold 4 squares diagonally to make triangles. Have students cut the squares along the fold line to make 8 triangles.

Alternatively, provide students with pre-cut triangles that are half-squares.

Have students put the triangles together to create pictures.

Learn

Have students put the triangles together to make the shapes in the textbook. Ask students what new shapes they have made.

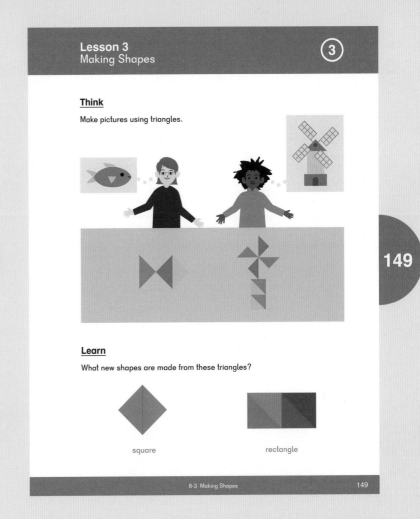

Do

4 Extend by providing additional right triangles and ask students about the relationship between the number of corners and number of outside lines that make up their new shape.

Challenge students to make a shape with either the least or the most corners.

Provide students with 2 circle pieces of paper and have them fold both in half and cut once. Fold the second paper circle in half again and cut to make quarters.

Allow students to make additional shapes with their pieces. They can glue the pieces onto paper to create pictures.

Exercise 3 • page 147

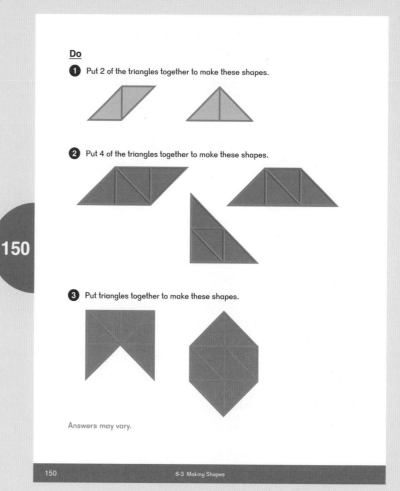

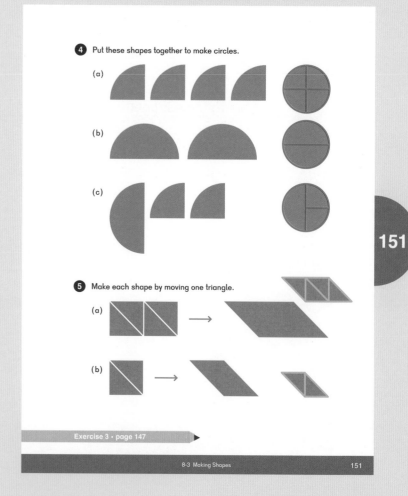

Objective

- Work with shapes.

After students complete the **Practice** in the textbook, have them continue to classify and discuss shapes. This skill lays the foundation for future studies in geometry.

Activities

▲ Tangrams

Materials: Tangram Pieces (BLM) copied onto cardstock, *Three Pigs, One Wolf, and Seven Magic Shapes* by Grace Maccarone, *Grandfather Tang's Story* by Ann Tompert

Tangrams can be used to develop problem-solving and logical thinking skills, visual-spatial awareness, creativity, and many mathematical concepts.

The book *Three Pigs, One Wolf, and Seven Magic Shapes* is an introduction to tangrams for younger students. For a greater challenge, try *Grandfather Tang's Story* by Ann Tompert.

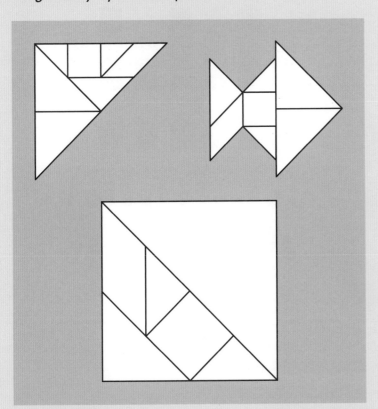

152

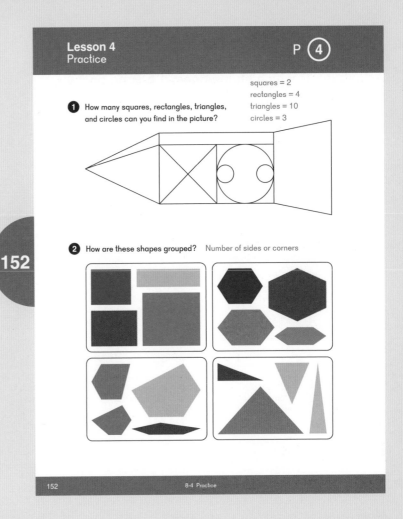

Lesson 4 Practice — P 4

1 How many squares, rectangles, triangles, and circles can you find in the picture?

squares = 2
rectangles = 4
triangles = 10
circles = 3

2 How are these shapes grouped? Number of sides or corners

152   8-4 Practice

▲ Attribute Train Game

Materials: Set of attribute blocks for each group of players (up to four or five players per group)

Player 1 places a shape (for example: a large blue triangle) to start the train. Player 2 looks for a shape that has only one attribute that differs from the first shape (in this example: a large red triangle).

Players continue to add to the train with shapes that have only one attribute that differs. In the example below, the next shape might be a small blue circle or triangle, a small yellow or red square or a large blue square.

▲ Which One Doesn't Belong?

Ask students which shape doesn't belong and why. There could be multiple reasons why each shape may not belong.

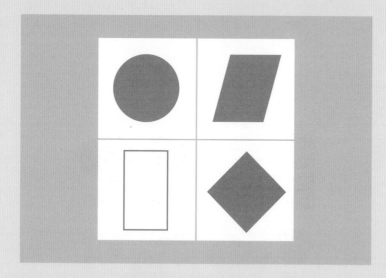

Brain Works

★ Count Them Up

How many squares and triangles?

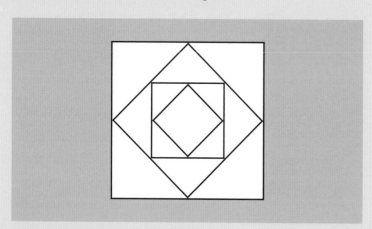

Answer:

4 squares

12 triangles

Exercise 4 · page 151

3 Make each shape using these shapes:

(a)

(b)

(c)

4 What shape is missing in the pattern?
What color is the missing shape?

(a)

(b)

(c)

5 How many figures are there in the repeating pattern?
What is the next figure?

(a)

2 figures

Next: tall rectangle

(b)

3 figures

Next:

(c)

5 figures

Next: square

(d)

4 figures

Next: cuboid

153

154

Exercise 4 · page 151

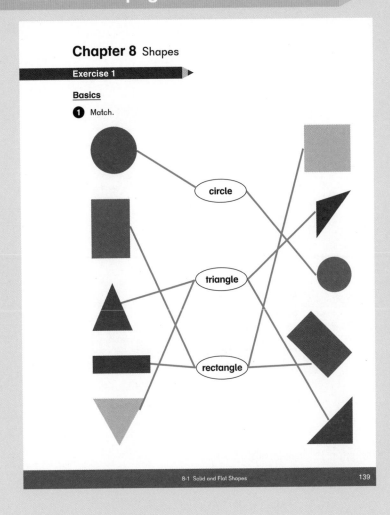

Chapter 8 Shapes

Exercise 1

Basics

1 Match.

circle

triangle

rectangle

8-1 Solid and Flat Shapes 139

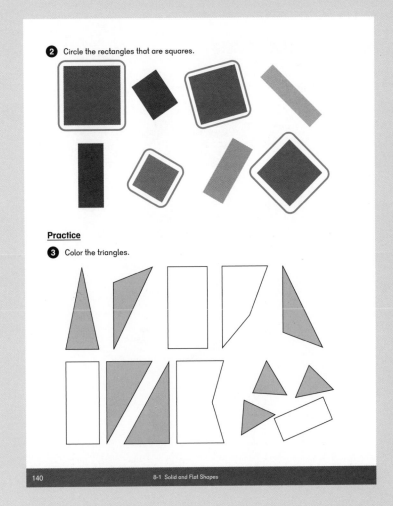

2 Circle the rectangles that are squares.

Practice

3 Color the triangles.

140 8-1 Solid and Flat Shapes

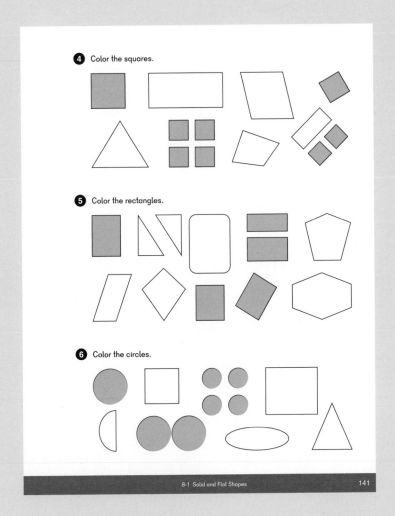

4 Color the squares.

5 Color the rectangles.

6 Color the circles.

8-1 Solid and Flat Shapes 141

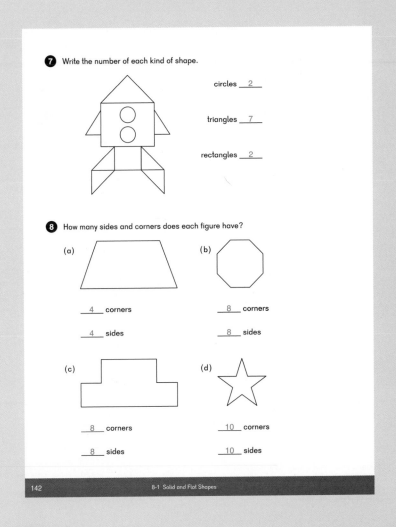

7 Write the number of each kind of shape.

circles _2_

triangles _7_

rectangles _2_

8 How many sides and corners does each figure have?

(a)

_4_ corners

_4_ sides

(b)

_8_ corners

_8_ sides

(c)

_8_ corners

_8_ sides

(d)

_10_ corners

_10_ sides

142 8-1 Solid and Flat Shapes

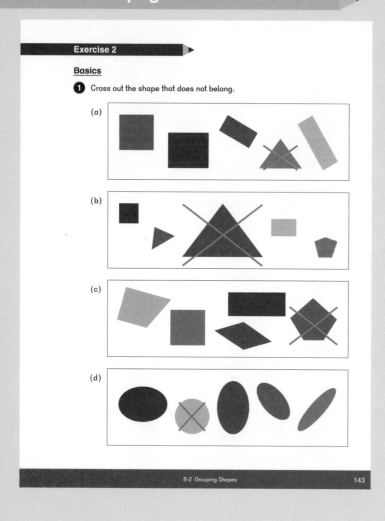

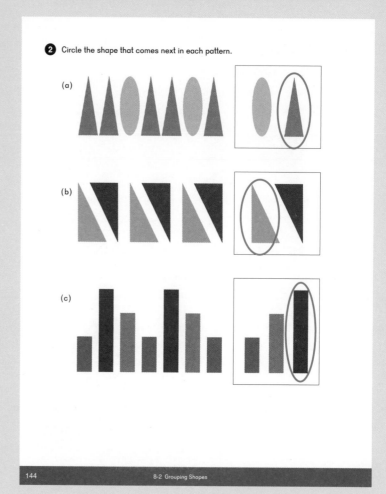

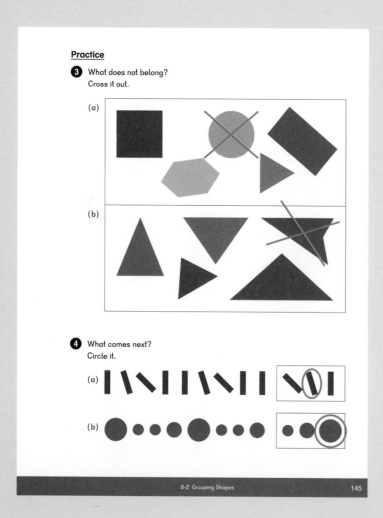

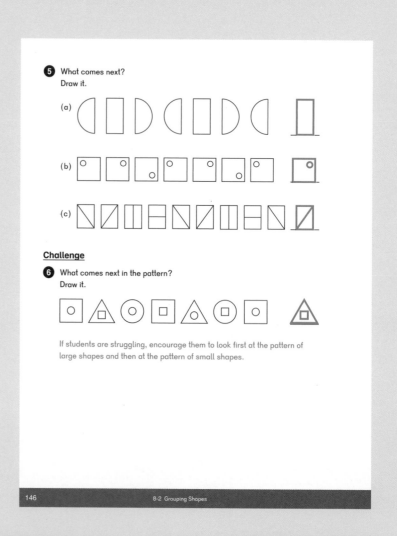

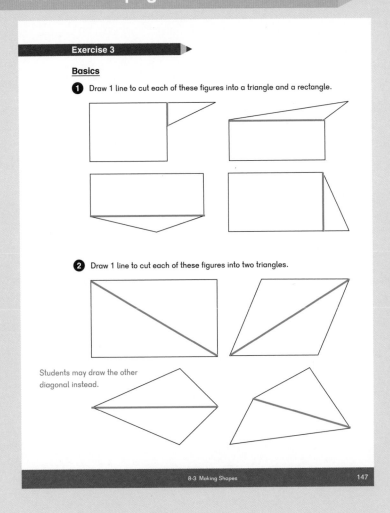

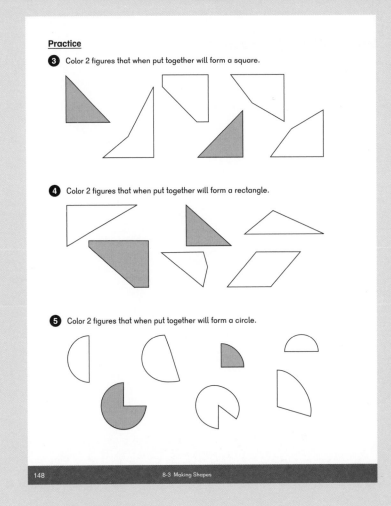

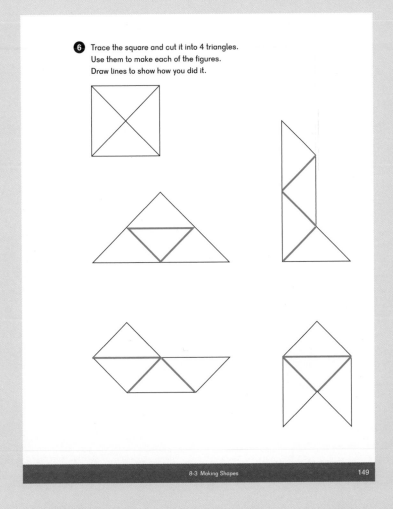

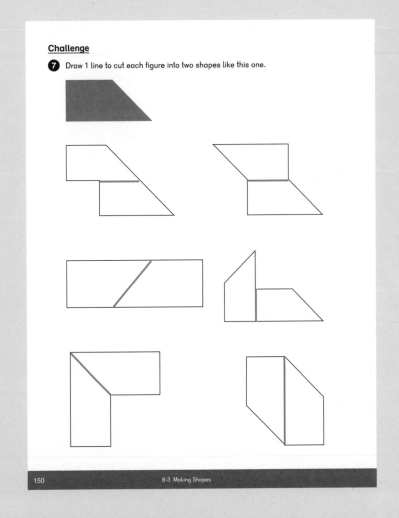

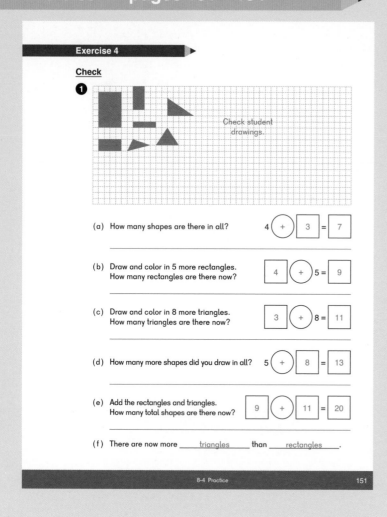

Exercise 4

Check

1

Check student drawings.

(a) How many shapes are there in all? 4 ⊕ 3 = 7

(b) Draw and color in 5 more rectangles.
How many rectangles are there now? 4 ⊕ 5 = 9

(c) Draw and color in 8 more triangles.
How many triangles are there now? 3 ⊕ 8 = 11

(d) How many more shapes did you draw in all? 5 ⊕ 8 = 13

(e) Add the rectangles and triangles.
How many total shapes are there now? 9 ⊕ 11 = 20

(f) There are now more ___triangles___ than ___rectangles___.

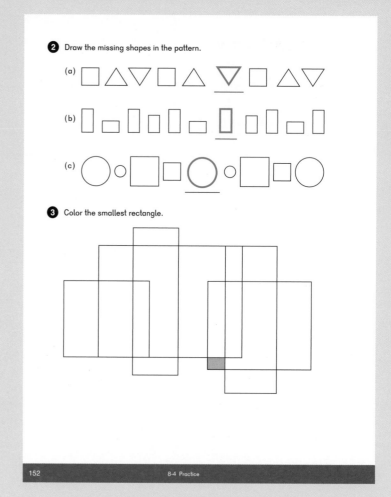

2 Draw the missing shapes in the pattern.

(a)

(b)

(c)

3 Color the smallest rectangle.

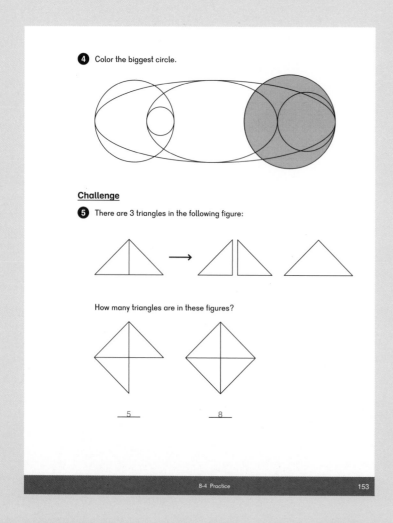

4 Color the biggest circle.

Challenge

5 There are 3 triangles in the following figure:

How many triangles are in these figures?

___5___   ___8___

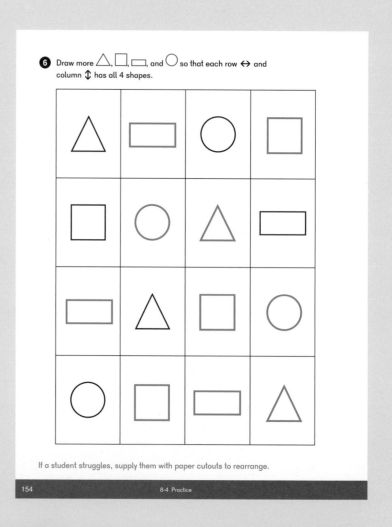

6 Draw more △, □, ▭, and ○ so that each row ↔ and column ↕ has all 4 shapes.

If a student struggles, supply them with paper cutouts to rearrange.

Suggested number of class periods: 4 – 5

Lesson		Page	Resources		Objectives
	Chapter Opener	p. 213	TB:	p. 155	Investigate position words.
1	Naming Positions	p. 214	TB: WB:	p. 156 p. 155	Understand the difference between ordinal numbers and cardinal numbers. Apply ordinal numbers to describe the position of an object in a line.
2	Word Problems	p. 217	TB: WB:	p. 160 p. 159	Solve word problems involving ordinal numbers.
3	Practice	p. 219	TB: WB:	p. 162 p. 163	Work with ordinal numbers.
	Review 2	p. 221	TB: WB:	p. 164 p. 167	Review content from Chapter 1 through Chapter 9.
	Workbook Solutions	p. 225			

Chapter 9: Ordinal Numbers focuses on position words and ordinal numbers.

In **Dimensions Math® Kindergarten**, students learned the names of positions from first to tenth. They also learned the terms left, right, front, back, top, and bottom.

In this chapter, students will learn the difference between ordinal numbers, which name position, and cardinal numbers, which name quantities. They should see that coloring 4 objects is different from coloring the fourth object.

The textbook pages for this chapter include more reading than prior chapters. Students should not be expected to read the problems independently.

Materials

- Index cards
- Two-color counters
- Art paper and markers, crayons, or paint
- Pattern Chains made in **Chapter 8: Lesson 2**
- Whiteboards

Storybooks

- Picture or board books showing several events
- *The Very Hungry Caterpillar* by Eric Carle
- *10 Little Rubber Ducks* by Eric Carle
- *20 Hungry Piggies* by Trudy Harris
- *Albert the Muffin-Maker* by Eleanor May
- *Henry the Fourth* by Stuart J. Murphy
- *Seven Blind Mice* by Ed Young

Letters Home

- Chapter 9 Letter

Chapter Opener

Objective

- Investigate position words.

Discuss the **Chapter Opener** with students with students.

Sample questions to ask:

- How many students are in line?
- Who is first in line? How do you know?
- What position is _____ in?

Have students line up in the classroom and discuss their positions in line, focusing on who is first and who is last. Ask students what we can say about the students between those in the first and last position.

Extend the **Chapter Opener** to a full lesson by reading one of the books listed on the previous page.

Chapter 9

Ordinal Numbers

155

Objectives

- Understand the difference between ordinal numbers and cardinal numbers.
- Apply ordinal numbers to describe the position of an object in a line.

Lesson Materials

- Index cards with 1st, 2nd, … 10th written on them

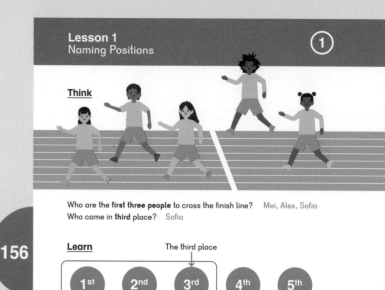

**Lesson 1
Naming Positions**

Think

Who are the **first three people** to cross the finish line?   Mei, Alex, Sofia
Who came in **third place**?   Sofia

Learn   The third place

| 1st | 2nd | 3rd | 4th | 5th |

| First | Second | Third | Fourth | Fifth |
| Mei | Alex | Sofia | Dion | Meg |

The first three people

156   9-1 Naming Positions

156

Think

Have 10 students line up single file in the classroom, all facing the same direction, with one student designated as the "line leader."

Ask students, "Who is first in line?"

Write "1st" and "first" on the board and tell students that they mean the same thing. Give the first student in line the "1st" card. Repeat for the student in second place.

Continue naming the positions and handing out the position cards through tenth place.

Collect the cards and have students face the rest of the class so they are still in line. "First" could be either end. Ask questions like:

- Which student is the second from the left?
- Who is second from the right?
- Who is fourth from the left?
- How many students are in front of the fourth student? (or behind the sixth or between the fourth and sixth in line)
- Who is fifth from last?
- Who are the last 5 students?

Learn

Have students discuss the places of each of the runners and relate the places to the position words: first, second, etc., through tenth.

Do

1 (a) — (k) Have students complete these problems verbally.

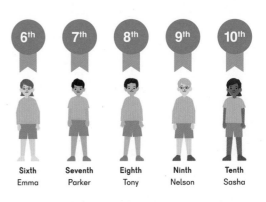

Sixth	Seventh	Eighth	Ninth	Tenth
Emma	Parker	Tony	Nelson	Sasha

9-1 Naming Positions   157

Do

1

(a) Which animal is **1st** in line?   turtle

(b) Which animals are the **first 5** in line?   turtle, ant, duck, dog, cat

(c) Which animal is **5th** in line?   cat

(d) Which animal is **5th from last**?   mountain goat

(e) Which animal is **2nd** in line?   ant

(f) Which animal is **7th** in line?   hermit crab

(g) In which position from the front is the mountain goat?   6th

(h) What position from the front is the last animal in line?   10th

(i) How many animals are **in front of the 8th** in line?   7

(j) How many animals are **behind the 8th** in line?   2

(k) How many animals are **between the 6th and 9th** in line?   2

158   9-1 Naming Positions

Activities

▲ I'm the nth.

Challenge students to find sequences and their place in them. For example:

- I am the third child, but the second oldest boy in my family.
- My cake won second prize at the fair.
- I am the fourth tallest in my family.
- My birthday is in the tenth month.

Have students illustrate their examples and share with the class.

▲ Is It Under... ?

Materials: 10 opaque cups, sharpie, 1 small object such as a counting bear, penny, linking cube, etc.

Prepare the game by writing 1st, 2nd, 3rd, ..., 10th on the cups as shown below so that they can be read when the cups are placed upside-down.

Partners take turns being the Finder and the Hider. The Hider lines the cups up in order. The Finder closes their eyes while the Hider places the object under one of the cups.

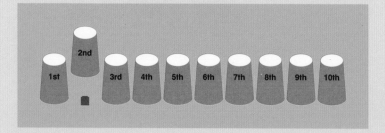

The Finder attempts to find the object by choosing cups. The Hider provides Clues to the correct cup. For example:

Finder: Is it under the fifth cup?
Hider: No, it is under a cup that is closer to first place.

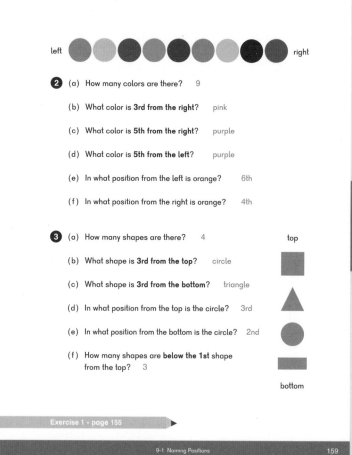

2 (a) How many colors are there?   9

(b) What color is **3rd from the right**?   pink

(c) What color is **5th from the right**?   purple

(d) What color is **5th from the left**?   purple

(e) In what position from the left is orange?   6th

(f) In what position from the right is orange?   4th

3 (a) How many shapes are there?   4

(b) What shape is **3rd from the top**?   circle

(c) What shape is **3rd from the bottom**?   triangle

(d) In what position from the top is the circle?   3rd

(e) In what position from the bottom is the circle?   2nd

(f) How many shapes are **below the 1st** shape from the top?   3

Exercise 1 • page 155

9-1 Naming Positions   159

Exercise 1 • page 155

Lesson 2 Word Problems

Objective

- Solve word problems involving ordinal numbers.

Lesson Materials

- Two-color counters, 10 per student

Think

Have students discuss the students in the line in the **Think** problem. Note that students cannot see the front of the line. Have them lay out counters of the same color for each student in line and flip one counter over to the other color to represent Mei.

Ask students:

- Where is Mei in the line?
- How many students are before Mei?
- How many are after her?

Learn

Have students look at the illustration under **Learn**. Ask them, "What equation would show how many students are in line altogether?"

Write "5 + 4 = 9" on the board.

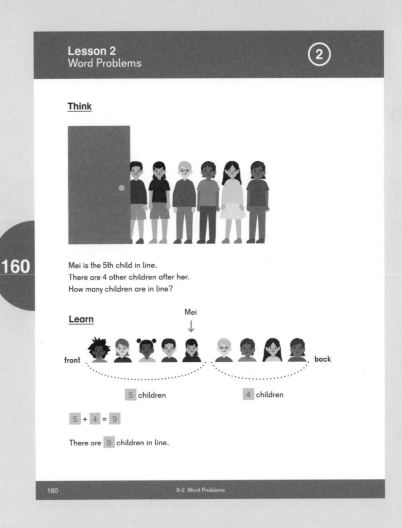

Do

Read the problems aloud.

❸ — **❹** Model with counters.

Students often struggle with ordinal numbers when they notice that there are nine positions from 2nd to 10th, but 10 − 2 = 8.

To help explain this oddity, remind students that someone in 2nd place would have to hop 8 places to reach 10, and they would not count the place they started on. 10 − 2 = 8. However, if they are counting how many places, including the one they are currently in, then there are nine such places.

Activity

▲ Pattern Chains

Materials: Pattern chains from **Chapter 8: Shapes**

Use the pattern chains from **Chapter 8: Lesson 2** on page 202 of this Teacher's Guide to have students discuss the shapes by their position. Ask questions like, "What is the third shape in the pattern?"

See if students can predict the sixth or tenth shape in the pattern.

Exercise 2 • page 159

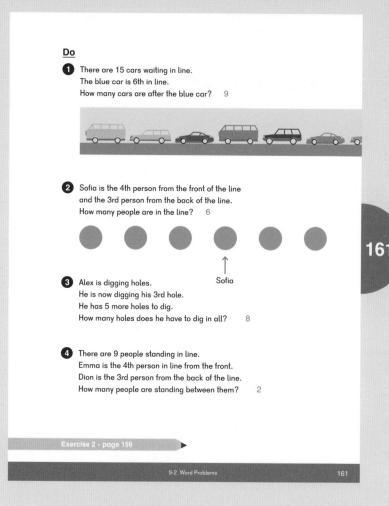

Objective

- Work with ordinal numbers.

Lesson Materials

- Two-color counters, if needed

Practice

After students complete the **Practice** in the textbook, have them continue to identify and use ordinal numbers in everyday school situations. (PE class, science experiments, etc.)

Activity

▲ **Story Sequencing**

Materials: Picture or board book (suggestions can be found on page 212 of this Teacher's Guide), art paper, markers, crayons, and/or paint

Choose a simple book with several events. For example, *The Very Hungry Caterpillar* by Eric Carle has simple pictures and a sequenced story.

Have students take turns describing what is happening in each picture. Ask students, "What happened *first* in the story?"

Students can illustrate the different stages in the story. Mix up the illustrations and have students re-sequence them based on when they happened in the story. Continue with additional questions about the second, third, and fourth events in the story.

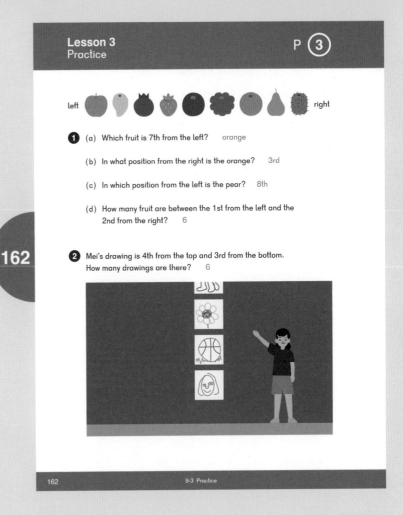

Review the pictures with students after they are put into order.

First, out popped a hungry caterpillar.
Second, he ate through one apple.
Third, he ate through two pears...

Brain Works

★ Order Clues

Can you put the friends in the correct order?

- Sofia and her friends are climbing the rock wall.
- Sofia is not second from the top.
- Alex is between Emma and Mei.
- Mei is at the top.
- Alex is third from the bottom.

Answer: Mei, Alex, Emma, Sofia

Exercise 3 • page 163

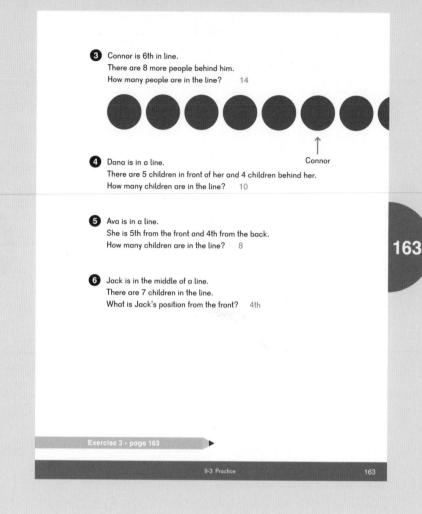

Review 2

Objective

- Review content from Chapter 1 through Chapter 9.

Use as necessary to review content and skills prior to end-of-book assessment.

Reviews are important not only to reinforce skills but also to assess formatively and provide remediation for students.

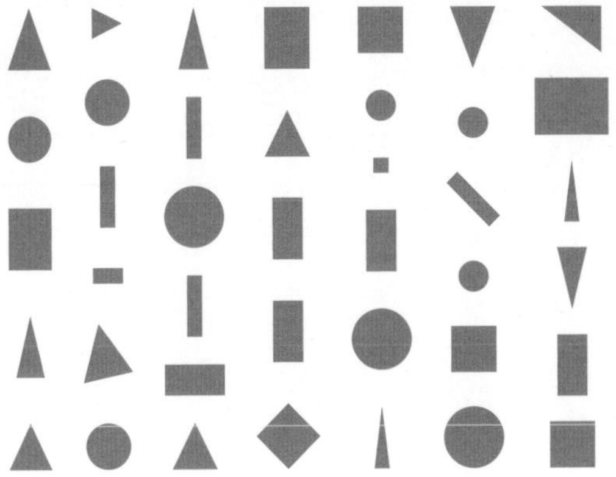

This review is cumulative and can be used as a basis for an assessment.

As these problems are review, students should be able to complete them without cards or counters.

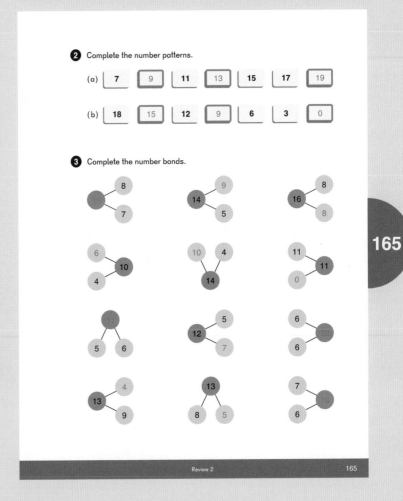

164

165

8 As Lisa has read all of pages 3 through 12, the subtraction equation 12 − 3 = 9 will not be sufficient to find the correct answer of 10 pages read.

Students could use counters or check this answer with pages from this book. It is possible they will see that the 12th page is included and solve this problem with 13 − 3 = 10.

---

4 Put in order from least to greatest.

| 6 | 11 | 14 | 15 | 18 |

(a) | 15 | 6 | 18 | 11 | 14 |

(b) | nineteen | twenty | eight | twelve | two |

| two | eight | twelve | nineteen | twenty |

(c) | 15 − 6 | 16 − 6 | 9 + 6 | 15 + 2 | 19 − 6 |

| 15 − 6 | 16 − 6 | 19 − 6 | 9 + 6 | 15 + 2 |

5 (a) 14 − 6 = 8   (b) 8 + 9 = 17

(c) 7 + 4 = 11   (d) 15 − 8 = 7

(e) 4 + 8 = 10 + 2   (f) 16 − 7 = 3 + 6

(g) 14 − 5 = 10 − 1   (h) 13 − 8 = 10 − 5

(i) 12 − 8 = 13 − 9   (j) 11 − 2 = 4 + 5

6 What comes next?

---

7 Write an equation for each and find the answer.

(a) Mari has 7 books.
She buys 5 more books.
How many books does she have now?
7 + 5 = 12

(b) Salim wants to read 15 books.
He has read 9 books so far.
How many more books does he still need to read?
15 − 9 = 6

(c) Tyler found 15 seashells.
3 of them are broken.
How many seashells are not broken?
15 − 3 = 12

(d) Holly gave away 3 seashells.
She now has 6 seashells.
How many seashells did she have at first?
3 + 6 = 9

(e) There are 14 dogs in an animal shelter.
7 of the dogs are adopted.
How many dogs are waiting to be adopted?
14 − 7 = 7

(f) 4 cats were adopted in the morning.
8 cats were adopted in the afternoon.
How many cats were adopted that day?
4 + 8 = 12

8 Lisa read from the beginning of the 3rd page to the end of the 12th page in a book.
How many pages did she read?
10 pages

Activity

● ### SET® Game

Materials: Cards from the SET® Game

This is a great age to teach students the game of SET®. Game sets can be purchased or cards can be found on the internet.

The emphasis is on finding attributes of shapes that are similar or different. The game can be played at multiple levels, making it enjoyable for all ages.

Why might these cards be a "Set?"

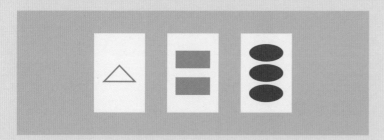

Exercise 4 • page 167

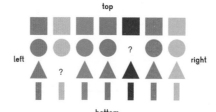

168

9 These figures are grouped into 4 rows and 7 columns.

top

left   ?   right

?

bottom

(a) How are they grouped into rows?   By shape

(b) How are they grouped into columns?   By color

(c) What are the shapes and colors of the missing figures?   Yellow triangle, purple circle

(d) What color is in the middle column?   blue

(e) What shapes are in the 3rd row from the bottom?   circles

(f) What color is in the 3rd column from the right?   purple

(g) What shape and color is 4th from the left and 4th from the top?   blue rectangle

(h) In what position is the green triangle?   3rd from the top, 3rd from the left (answers may vary)

Exercise 4 • page 167

168   Review 2

Notes

Chapter 9 Ordinal Numbers

Exercise 1

Basics

1. (a) Match.

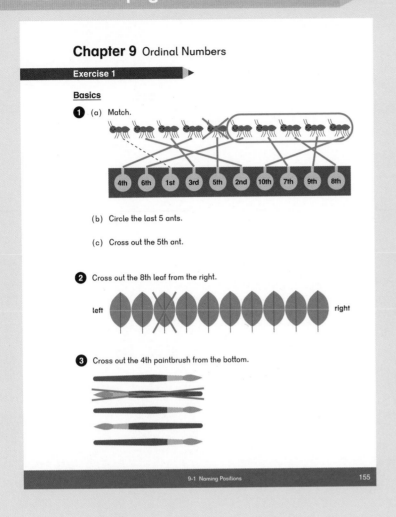

(b) Circle the last 5 ants.

(c) Cross out the 5th ant.

2. Cross out the 8th leaf from the right.

left   right

3. Cross out the 4th paintbrush from the bottom.

9-1 Naming Positions   155

Practice

4. Color the 4th banana from the left.

left   right

5. Color the first 4 balloons.

1st

6. Circle the 3rd bead from the knot.

7. The lollipop in the middle is __5th__ from the right.

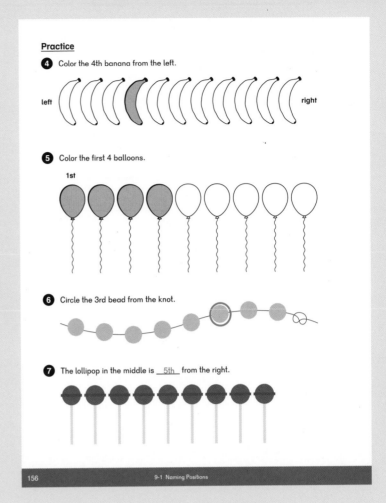

156   9-1 Naming Positions

8. The milk is __6th__ from the left.

How many drinks are to the right of the juice box? __7__

9. Color the 11th flower.
How many flowers are between the 9th and the 13th flowers? __3__

1st   9th   3   13th

10. (a) Circle the 3rd item from the top.

(b) The glue stick is __4th__ from the top.

(c) The pencil is __5th__ from the bottom.

(d) There are __2__ items above the 3rd item from the top.

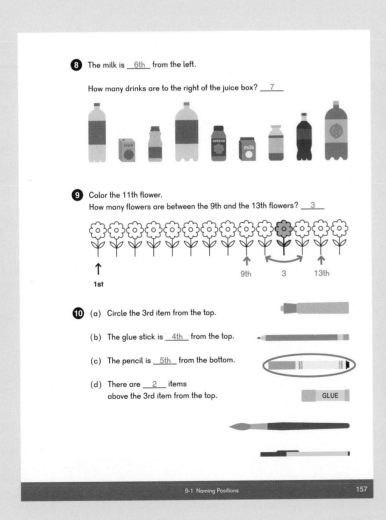

9-1 Naming Positions   157

11. Match.

ninth   2nd

fourteenth   3rd

second   5th

fifth   20th

eighth   1st

third   9th

twelfth   8th

twentieth   10th

first   14th

tenth   12th

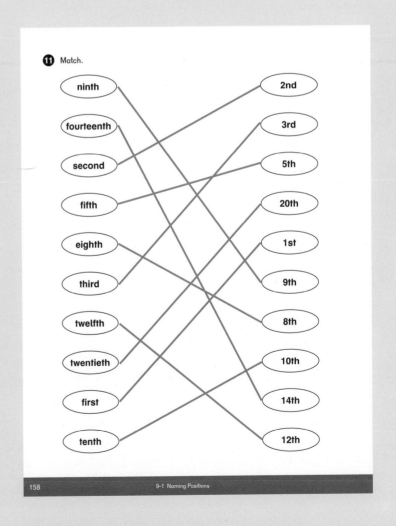

158   9-1 Naming Positions

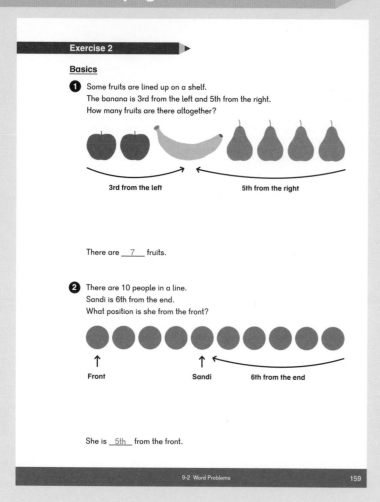

Exercise 2

Basics

1 Some fruits are lined up on a shelf.
The banana is 3rd from the left and 5th from the right.
How many fruits are there altogether?

3rd from the left   5th from the right

There are __7__ fruits.

2 There are 10 people in a line.
Sandi is 6th from the end.
What position is she from the front?

Front   Sandi   6th from the end

She is __5th__ from the front.

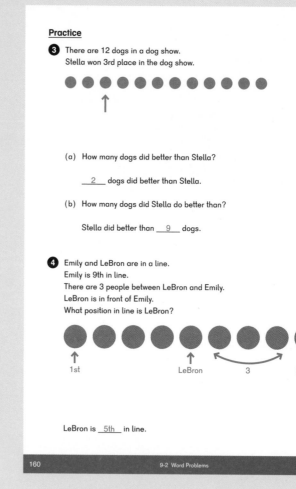

Practice

3 There are 12 dogs in a dog show.
Stella won 3rd place in the dog show.

(a) How many dogs did better than Stella?

__2__ dogs did better than Stella.

(b) How many dogs did Stella do better than?

Stella did better than __9__ dogs.

4 Emily and LeBron are in a line.
Emily is 9th in line.
There are 3 people between LeBron and Emily.
LeBron is in front of Emily.
What position in line is LeBron?

1st   LeBron   3   Emily

LeBron is __5th__ in line.

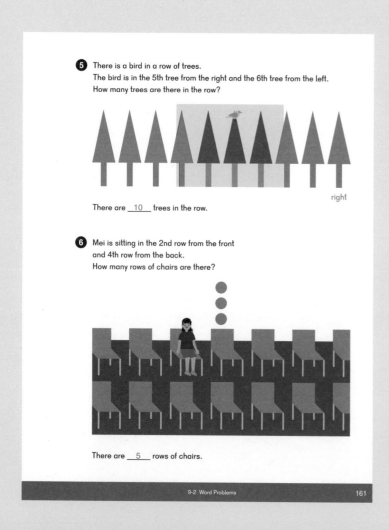

5 There is a bird in a row of trees.
The bird is in the 5th tree from the right and the 6th tree from the left.
How many trees are there in the row?

right

There are __10__ trees in the row.

6 Mei is sitting in the 2nd row from the front
and 4th row from the back.
How many rows of chairs are there?

There are __5__ rows of chairs.

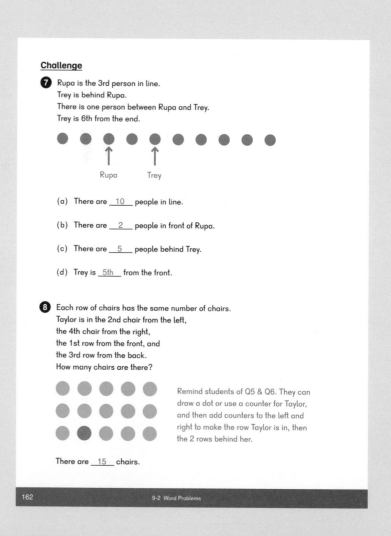

Challenge

7 Rupa is the 3rd person in line.
Trey is behind Rupa.
There is one person between Rupa and Trey.
Trey is 6th from the end.

Rupa   Trey

(a) There are __10__ people in line.

(b) There are __2__ people in front of Rupa.

(c) There are __5__ people behind Trey.

(d) Trey is __5th__ from the front.

8 Each row of chairs has the same number of chairs.
Taylor is in the 2nd chair from the left,
the 4th chair from the right,
the 1st row from the front, and
the 3rd row from the back.
How many chairs are there?

Remind students of Q5 & Q6. They can draw a dot or use a counter for Taylor, and then add counters to the left and right to make the row Taylor is in, then the 2 rows behind her.

There are __15__ chairs.

Exercise 3

Check

1

(a) Circle the 1st snowman from the right.

(b) Draw a hat on the 4th snowman from the left.

(c) Draw noses on 3 of the snowmen.   Which 3 can vary.

(d) Draw buttons on the 3rd snowman from the right.

(e) The biggest snowman is __6th__ from the left.

(f) The smallest snowman is __2nd__ from the left.

(g) How many snowmen are between the 3rd and 6th snowmen? __2__

(h) Mariya made some more snowmen.
The 6th snowman from the left is now in the middle.
How many snowmen are there now?

There are now __11__ snowmen.

2 9 pies are lined up in a row.
Alex picked the pie in the middle of the row.
What position is the pie he picked from the left?

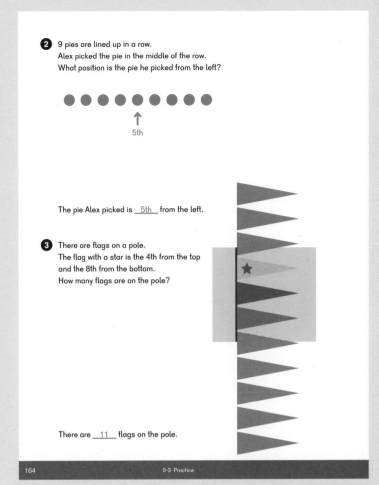

The pie Alex picked is __5th__ from the left.

3 There are flags on a pole.
The flag with a star is the 4th from the top
and the 8th from the bottom.
How many flags are on the pole?

There are __11__ flags on the pole.

4

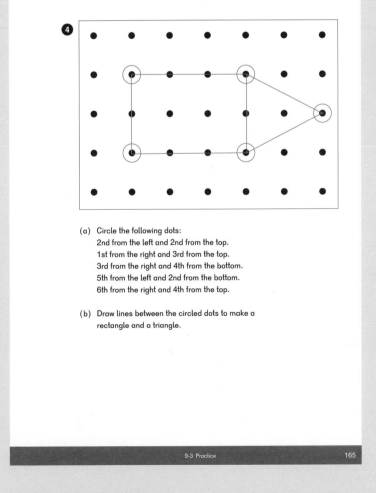

(a) Circle the following dots:
2nd from the left and 2nd from the top.
1st from the right and 3rd from the top.
3rd from the right and 4th from the bottom.
5th from the left and 2nd from the bottom.
6th from the right and 4th from the top.

(b) Draw lines between the circled dots to make a rectangle and a triangle.

5 Circle the shape that will be 10th in the pattern.

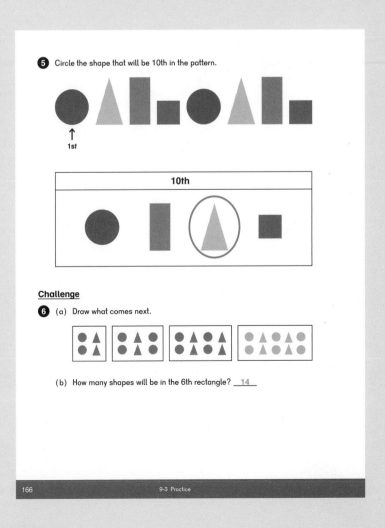

Challenge

6 (a) Draw what comes next.

(b) How many shapes will be in the 6th rectangle? __14__

Teacher's Guide 1A Chapter 9

Exercise 4

Check

1 (a) Write the numbers in order from least to greatest.

| 13 | 11 | 7 | 15 | 9 |

| 7 | 9 | 11 | 13 | 15 | 17 |
1st

(b) In the 6th box, write the number that comes next in the pattern.

(c) Add the 1st and 3rd number in the pattern.

$$7 + 11 = 18$$

(d) Subtract the 1st number from the 6th number in the pattern.

$$17 - 7 = 10$$

(e) Add the first 2 numbers in the pattern.

$$7 + 9 = 16$$

(f) Subtract the 2nd number from the 5th number in the pattern.

$$15 - 9 = 6$$

2

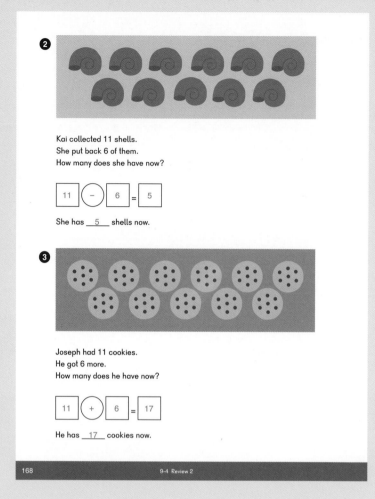

Kai collected 11 shells.
She put back 6 of them.
How many does she have now?

$$11 - 6 = 5$$

She has __5__ shells now.

3

Joseph had 11 cookies.
He got 6 more.
How many does he have now?

$$11 + 6 = 17$$

He has __17__ cookies now.

4

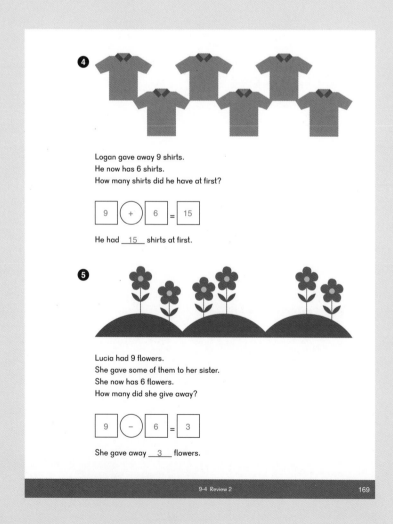

Logan gave away 9 shirts.
He now has 6 shirts.
How many shirts did he have at first?

$$9 + 6 = 15$$

He had __15__ shirts at first.

5

Lucia had 9 flowers.
She gave some of them to her sister.
She now has 6 flowers.
How many did she give away?

$$9 - 6 = 3$$

She gave away __3__ flowers.

6 Cross out any that are less than 15.

| 7 + 9 | 8 + 3 | 19 – 3 | 11 + 3 | 16 + 5 |

7 Cross out any that are more than 8.

| 5 + 6 | 2 + 5 | 17 – 8 | 11 – 4 | 17 – 6 |

8 Cross out the one that does not belong.

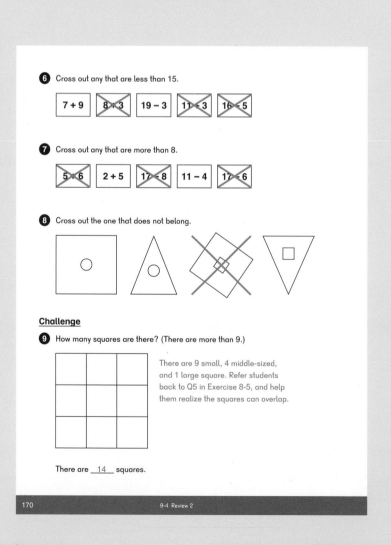

Challenge

9 How many squares are there? (There are more than 9.)

There are 9 small, 4 middle-sized, and 1 large square. Refer students back to Q5 in Exercise 8-5, and help them realize the squares can overlap.

There are __14__ squares.

Blackline Masters for 1A

All Blackline Masters used in the guide can be downloaded from dimensionsmath.com.

This lists BLM used in the **Think** and **Learn** sections.

BLMs used in **Activities** are listed in the Activity Materials within each chapter.

Blank Double Ten-frames	**Chapter 5:** Lesson 2, Lesson 4 **Chapter 6:** Lesson 1, Lesson 2, Lesson 3 **Chapter 7:** Lesson 1, Lesson 2, Lesson 3
Blank Ten-frame	**Chapter 1:** Lesson 1, Lesson 3 **Chapter 5:** Lesson 5, Lesson 6
Number Bond Story Template	**Chapter 2:** Chapter Opener, Lesson 3 **Chapter 3:** Lesson 1
Number Cards	**Chapter 1:** Lesson 2, Lesson 3 **Chapter 2:** Lesson 1, Lesson 6 **Chapter 3:** Lesson 1 **Chapter 4:** Lesson 9 **Chapter 5:** Lesson 1, Lesson 3 **Chapter 6:** Lesson 4 **Chapter 7:** Lesson 5
Number Word Cards	**Chapter 1:** Lesson 2 **Chapter 5:** Lesson 1
Shapes Worksheet	**Chapter 8:** Lesson 2
Spinner	**Chapter 3:** Lesson 2 **Chapter 4:** Lesson 9
Spinner 11 – 20	**Chapter 6:** Lesson 4
Ten-frame Cards	**Chapter 1:** Lesson 2 **Chapter 2:** Lesson 1, Lesson 2, Lesson 3, Lesson 4, Lesson 5 **Chapter 3:** Lesson 7 **Chapter 4:** Lesson 9 **Chapter 5:** Lesson 1

Notes